Principles of Irrigation

Principles of Irrigation

Nancy Faulkner

Syrawood Publishing House
New York

Published by Syrawood Publishing House,
750 Third Avenue, 9th Floor,
New York, NY 10017, USA
www.syrawoodpublishinghouse.com

Principles of Irrigation
Nancy Faulkner

International Standard Book Number: 978-1-64740-057-6 (Hardback)

Cataloging-in-Publication Data

Principles of irrigation / Nancy Faulkner.
p. cm.
Includes bibliographical references and index.
ISBN 978-1-64740-057-6
1. Irrigation. 2. Water in agriculture. 3. Water-supply, Agricultural.
4. Crops and water. I. Faulkner, Nancy.
S613 .P75 2022
631.587--dc23

Table of Contents

Preface

This book has been written, keeping in view that students want more practical information. Thus, my aim has been to make it as comprehensive as possible for the readers. I would like to extend my thanks to my family and co-workers for their knowledge, support and encouragement all along.

The application of water to plants at necessary intervals is known as irrigation. It is used to ensure the proper growth of agricultural crops and maintain landscapes. Irrigation plays a key role in re-vegetating disturbed soils in dry areas and during the times of less than average rainfall. Irrigation systems are also used for dust suppression, cooling livestock and protecting crops from frost. Some of the common methods of irrigation are surface irrigation, sprinkler irrigation and subsurface irrigation. There are various sources of water which can be used for irrigation such as groundwater, surface water and treated wastewater. The topics covered in this extensive textbook deal with the core aspects of irrigation. It is a valuable compilation of topics, ranging from the basic to the most complex methods in this field. Those in search of information to further their knowledge will be greatly assisted by this book.

A brief description of the chapters is provided below for further understanding:

Chapter - Introduction

The supply of water at definite intervals to land and crops for supporting their growth is known as irrigation. It includes the practices such as bioirrigation, biosalinity, fertigation, etc. This is an introductory chapter which will briefly introduce all these practices of irrigation as well as their impacts.

Chapter - Types and Methods of Irrigation

Several types of methods are used in irrigation such as check basin method, furrow irrigation, sprinkler irrigation, drip irrigation, pot irrigation, tidal irrigation, surface irrigation, micro-irrigation, sprinkler irrigation etc. The topics elaborated in this chapter will help in gaining a better perspective about these traditional and modern methods of irrigation.

Chapter - Water Sources for Irrigation

Various water sources that are used for irrigation are ground water, tube well, surface water, irrigation canals and tanks. Some of the non-conventional sources are also used for irrigation including desalinated water, spate irrigation and rain harvesting. This chapter closely examines these water sources of irrigation to provide an extensive understanding of the subject.

Chapter - Irrigation Management

The practice of monitoring and managing the rate, timing and volume of water according to the crop needs is referred to as irrigation water management. Irrigation scheduling, irrigation system maintenance, etc. are some of the methods used for irrigation management. This chapter has been carefully written to provide an easy understanding of irrigation water management.

Chapter – Irrigation Machinery

There are many important machines that are used in irrigation such as soil moisture sensors, irrigation controllers, irrigation system heads, disc filters, irrigation sprinklers and sprinkler system timers. The diverse applications of these irrigation machines have been thoroughly discussed in this chapter.

Chapter – Irrigation and the Environment

The environmental impacts of irrigation include changes in the quantity and quality of soil and water, and instability in the atmosphere. It also affects the natural and social conditions of river basins. The topics elaborated in this chapter will help in gaining a better perspective about these environmental impacts of irrigation.

Chapter – Waterlogging

The saturation of soil with water is referred to as waterlogging. Various methods are used to prevent water logging such as land grading, tree planting, reduction in unnecessary percolation of rainfall, efficient drainage system, in-field water management, etc. This chapter closely examines the key concepts of waterlogging to provide an extensive understanding of the subject.

Nancy Faulkner

1

Introduction

The supply of water at definite intervals to land and crops for supporting their growth is known as irrigation. It includes the practices such as bioirrigation, biosalinity, fertigation, etc. This is an introductory chapter which will briefly introduce all these practices of irrigation as well as their impacts.

IRRIGATION

Irrigation is the artificial application of water to land for the purpose of agricultural production. Effective irrigation will influence the entire growth process from seedbed preparation, germination, root growth, nutrient utilisation, plant growth and regrowth, yield and quality.

The key to maximising irrigation efforts is uniformity. The producer has a lot of control over how much water to supply and when to apply it but the irrigation system determines uniformity. Deciding which irrigation systems is best for your operation requires a knowledge of equipment, system design, plant species, growth stage, root structure, soil composition, and land formation. Irrigation systems should encourage plant growth while minimising salt imbalances, leaf burns, soil erosion, and water loss. Losses of water will occur due to evaporation, wind drift, run-off and water (and nutrients) sinking deep below the root zone.

The Value of Irrigation

Irrigation allows primary producers:

- To grow more pastures and crops.
- To have more flexibility in their systems/operations as the ability to access water at times

when it would otherwise be hard to achieve good plant growth (due to a deficit in soil moisture) is imperative. Producers can then achieve higher yields and meet market/seasonal demands especially if rainfall events do no occur.

- To produce higher quality crops/pastures as water stress can dramatically impact on the quality of farm produce.
- To lengthen the growing season (or in starting the season at an earlier time).
- To have 'insurance' against seasonal variability and drought.
- To stock more animals per hectare and practice tighter grazing management due to the reliability of pasture supply throughout the season.
- To maximise benefits of fertiliser applications. Fertilisers need to be 'watered into' the ground in order to best facilitate plant growth.
- To use areas that would otherwise be 'less productive'. Irrigation can allow farmers to open up areas of their farms where it would otherwise be 'too dry' to grow pasture/crops. This also gives them the capability to carry more stock or to conserve more feed.
- To take advantage of market incentives for unseasonal production.
- To have less reliance on supplementary feeding (grain, hay) in grazing operations due to the more consistent supply & quality of pastures grown under irrigation.
- To improve the capital value of their property. Since irrigated land can potentially support higher crops, pasture and animal production, it is considered more valuable. The value of the property is also related to the water licensing agreements or 'water right'.
- To cost save/obtain greater returns. The cost benefits from the more effective use of fertilisers and greater financial benefits as a result of more effective agricultural productivity (both quality and quantity) and for 'out of season' production are likely.

Problems

While irrigation has provided a number of important benefits the potential drawbacks of over/under watering include:

Under-watering

- Loss in market value through yield reduction.
- Reduction in fruit size and quality.

Over-watering

- Unwanted vegetative growth.
- Losses of valuable water to the watertable.

- Irrigation water travelling over soil can cause erosion. The excessive displacement of the top soil can also affect soil fertility (and hence crop yields), it may also clog drainage ditches and streams (silting), harm aquatic habitats, foul waters used for recreational activities, and increases the need for water treatments.
- Irrigation can cause pesticides, pathogens and weeds to spread during irrigation.
- Cause runoff.
- Increased operational costs (labour, pumping, cost of water).
- Leaching of nutrients (eg. salt, phosphorus) may lead to algal growth, salinity an nitrate build ups (poisoning) elsewhere in the catchment.
- Downgraded product quality and reduced yield.
- Higher operational costs for the producer (hence, reduced profits).
- Pressue on water resources with the Increasing demand for water use by urban dwellers.

BIOIRRIGATION

Bioirrigation refers to the process of benthic organisms flushing their burrows with overlying water. The exchange of dissolved substances between the porewater and overlying seawater that results is an important process in the context of the biogeochemistry of the oceans.

Coastal aquatic environments often have organisms that destabilize Sediment. They change the physical state of the sediment. Thus improving the conditions for other organisms and themselves. These organisms often also cause Bioturbation, which is commonly used interchangeably or in reference with bioirrigation.

Bioirrigation works as two different processes. These processes are known as particle reworking and ventilation, which is the work of benthic macro-invertebrates (usually ones that burrow). This particle reworking and ventilation is caused by the organisms when they feed (faunal feeding), defecate, burrow, and respire.

Bioirrigation is responsible for a large amount of oxidative transport and has a large impact on biogeochemical cycles.

Bioirrigation's Role in Elemental Cycling

Bioirrigation is a main component in element cycling. Some of these elements include: Magnesium, Nitrogen, Calcium, Strontium, Molybdenum, and Uranium. Other elements are only displaced at certain steps in the bioirrigation process. Aluminium, Iron, Cobalt, Copper, Zinc, and Cerium are all affected at the start of the process, when the larvae begins to dig into the sediment. While Manganese, Nickel, Arsenic, Cadmium and Caesium were all mobilized slightly after the burrowing process.

Challenges to Studying Bioirrigation

When trying to describe this biologically driven dynamic process, scientists have not been able to develop a 3D image of the process yet.

New Mechanisms to Study Bioirrigation

There is a hybrid medical imaging technique using a position emission tomography/computed tomography (PET/CT) to measure the ventilation and visualize the pore water advection that is caused by the organisms in 4D imaging.

Ecological Importance of Bioirrigation

When coastal ecosystems do not have bioirrigating organisms, like lugworms, it results in a lot of sedimentary problems. Some of these problems include clogging of the sediment with organic-rich fine particles and a drastic decrease in sediment permeability. It also makes it so the oxygen cannot penetrate deeply into the sediment and there is accumulation of reduced mineralized products in pore water. These problems disrupt the foundations of a coastal ecosystem.

Economic Impacts

Two organisms that contribute to the bioturbation of soil are Nephtys caeca (Fabricius) and Nereis virens (Sars) annelidae. They dig, bioirrigate, and feed in the sediment and they homogenize the particles found in the sediment when they partake in these activities because of their erratic movements. The bioirrigation generated by these organisms modifies the distribution of dinoflagellate cysts in the sedimentary column. They either bury them or raise them back to the surface, keeping them rotating. One of the most important dinoflagellates that these organisms help distribute is called noxious microalgae and it is responsible for the formation of toxic red tides. These red tides poison mollusks and crustaceans which results in very important economic losses in the fishing industry.

BIOSALINITY

Biosalinity is the study and practice of using saline (salty) water for irrigating agricultural crops. Many arid and semi-arid areas actually do have sources of water, but the available water is usually brackish (0.5–5g/L salt) or saline (30–50g/L salt). The water may be present in underground aquifers or as seawater along coastal deserts. With traditional farming practices, saline water results in soil salinization, rendering it unfit for raising most crop plants. Indeed, many arid and semi-arid areas were simply considered unsuitable for agriculture, and agricultural development of these areas was not systematically attempted until the second half of the 20th century.

Research in biosalinity includes studies of the biochemical and physiological mechanisms of salt tolerance in plants, breeding and selection for salt tolerance (halotolerance), discovery of periods in a crop plant's life cycle when it may be less sensitive to salt, use of saline irrigation water to increase desirable traits (such as sugar concentration in a fruit) or to control the ripening process, study of the interaction between salinity and soil properties, and development of naturally salt-tolerant plant species (halophytes) into useful agricultural crops.

When properly applied (watering well in excess of evapotranspiration, maintaining soil structure for excellent drainage), brackish-water irrigation does not result in increased salinization of the soil. Sometimes this means that farmers have to add extra water after a rainstorm, to carry salts back down to below the root zone.

FERTIGATION

Fertigation is the injection of fertilizers, used for soil amendments, water amendments and other water-soluble products into an irrigation system.

Fertigation is related to chemigation, the injection of chemicals into an irrigation system. The two terms are sometimes used interchangeably however chemigation is generally a more controlled and regulated process due to the nature of the chemicals used. Chemigation often involves insecticides herbicides, and fungicides, some of which pose health threat to humans, animals, and the environment.

Uses

Fertigation is practiced extensively in commercial agriculture and horticulture. Fertigation is also increasingly being used for landscaping as dispenser units become more reliable and easier to use. Fertigation is used to add additional nutrients or to correct nutrient deficiencies detected in plant tissue analysis. It is usually practiced on the high-value crops such as vegetables, turf, fruit trees, and ornamentals.

Commonly used Nutrients

Most plant nutrients can be applied through irrigation systems. Nitrogen is the most commonly used plant nutrient. Naturally occurring nitrogen (N_2) is a diatomic molecule which makes up approximately 80% of the earth's atmosphere. Most plants cannot directly consume diatomic nitrogen, therefore nitrogen must be contained as a component of other chemical substances which plants can consume. Commonly, anhydrous ammonia, ammonium nitrate, and urea are used as bioavailable sources of nitrogen. Other nutrients needed by plants include phosphorus and potassium. Like nitrogen, plants require these substances to live but they must be contained in other chemical substances such as monoammonium phosphate or diammonium phosphate to serve as bioavailable nutrients. A common source of potassium is muriate of potash which is chemically potassium chloride. A soil fertility analysis is used to determine which of the more stable nutrients should be used.

Advantages

The benefits of fertigation methods over conventional or drop-fertilizing methods include:

- Increased nutrient absorption by plants.
- Accurate placement of nutrient, where the water goes the nutrient goes as well.
- Ability to "microdose", feeding the plants just enough so nutrients can be absorbed and are not left to be washed down to stormwater next time it rains.
- Reduction of fertilizer, chemicals, and water needed.
- Reduced leaching of chemicals into the water supply.
- Reduced water consumption due to the plant's increased root mass's ability to trap and hold water.
- Application of nutrients can be controlled at the precise time and rate necessary.
- Minimized risk of the roots contracting soil borne diseases through the contaminated soil.
- Reduction of soil erosion issues as the nutrients are pumped through the water drip system. Leaching is decreased often through methods used to employ fertigation.

Disadvantages

Concentration of the solution may decrease as the fertilizer dissolves, this depends on equipment selection. If poorly selected may lead to poor nutrient placement:

- The water supply for fertigation is to be kept separate from the domestic water supply to avoid contamination.
- Possible pressure loss in the main irrigation line.
- The process is dependent on the water supply's non-restriction by drought rationing.

Methods Used

- Drip irrigation: Less wasteful than sprinklers. It is not only more efficient for fertilizer usage, but can also be for maximizing nutrient uptake in plants like cotton. Drip irrigation using fertigation can also increase yield and quality of fruit and flowers, especially in subsurface drip systems rather than above surface drip tape.

- Sprinkler systems: Increases leaf and fruit quality.
- Continuous application: Fertilizer is supplied at a constant rate.
- Three-stage application: Irrigation starts without fertilizers. Fertilizers are applied later in the process.
- Proportional application: Injection rate is proportional to water discharge rate.
- Quantitative application: Nutrient solution is applied in a calculated amount to each irrigation block.
- Other methods of application include the lateral move, the traveler gun, and solid set systems.

System Design

Fertigation assists distribution of fertilizers for farmers. The simplest type of fertigation system consists of a tank with a pump, distribution pipes, capillaries, and a dripper pen.

All systems should be placed on a raised or sealed platform, not in direct contact with the earth. Each system should also be fitted with chemical spill trays.

Because of the potential risk of contamination in the potable (drinking) water supply, a backflow prevention device is required for most fertigation systems. Backflow requirements may vary greatly. Therefore, it is very important to understand the proper level of backflow prevention required by law. In the United States, the minimum backflow protection is usually determined by state regulation. Each city or town may set the level of protection required.

IRRIGATION INFORMATICS

Irrigation informatics is a newly emerging academic field that is a cross-disciplinary science using informatics to study the information flows and data management related to irrigation. The field is one of many new informatics sub-specialities that uses the science of information, the practice of information processing, and the engineering of information systems to advance a biophysical science or engineering field.

Agricultural productivity increases are eagerly sought by governments and industry, spurred by the realisation that world food production must double in the 21st century to feed growing populations and that as irrigation makes up 36% of global food production, but that new land for irrigation growth is very limited, irrigation efficiency must increase. Since irrigation science is a mature and stable field, irrigation researchers are looking to cross-disciplinary science to bring about production gains and informatics is one such science along with others such as social science. Much of the driver for work in the area of irrigation informatics is the perceived success of other informatics fields such as health informatics.

Irrigation informatics is very much a part of the wider research into irrigation wherever information technology or data systems are used, however the term *informatics* is not always used to describe research involving computer systems and data management so that *information science* or *information technology* may alternatively be used.

Recent work in the general area of irrigation informatics has mentioned the exact phrase "Irrigation Informatics" with at least one publication in scientific conference proceedings using it in its title.

Current Implementations

Meteorological informatics, as with all informatics, are increasingly being used to handle the growing volumes of data that are available from sensors, remote sensing and scientific models. The Australian Bureau of Meteorology has recently implemented an XML data format, known as the Water Data Transfer Format (WDTF) and standard to be used by Australian government agencies and meteorological data suppliers when delivering data to the Bureau. This format includes specifications for evapotranspiration and other weather parameters that are useful for irrigation and may be used through implementations of irrigation informatics.

WICKING BEDS

Wicking beds are a unique and increasingly popular way to grow vegetables. They are self-contained raised beds with built-in reservoirs that supply water from the bottom up – changing how, and how much, you water your beds.

How Wicking Beds Work?

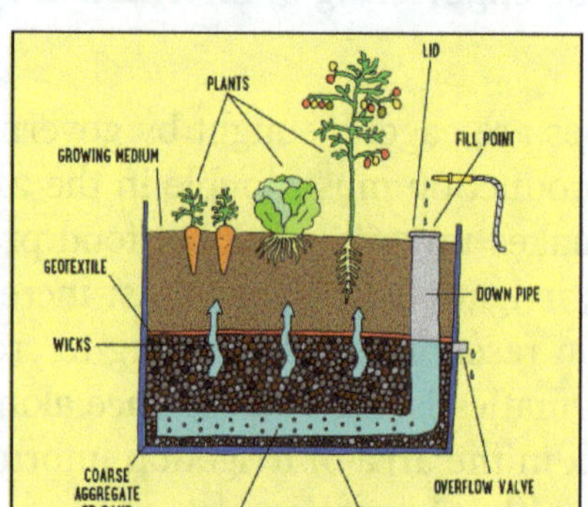

A wick works through capillary action – the same force you observe when you dip a piece of tissue paper partially into a glass of water and watch the water climb the paper. Wicking occurs in many materials; cotton, wool, geo-textile, soil, gravel and even wood to some degree. Every material has different wicking properties which you can test by placing that material into a glass of water and watching the water "climb" up. When one end of the wick is saturated and the other end is dry, it creates a moisture gradient, which drives the wick until the gradient no longer exists or you run out of water. With the earth box, one of the more popular examples in North America, the soil is suspended above the reservoir with wicks dangling into the reservoir pulling up moisture. As the plants use the moisture in the soil, it creates a moisture gradient (the soil is drier than the reservoir) which drives moister through the wick into the soil.

Advantages of Wicking Beds

Wicking beds have a lot of advantages over standard raised beds and in-grown swale-based gardens:

- They are water-efficient: Watering from the bottom up prevents evaporation of surface water (which occurs when you water beds from the top).
- They are self-watering: Wicking beds are an especially great system to use in community gardens because they save people from driving every day during hot weeks to water their beds. A full wicking bed should irrigate itself for about a week.
- They can be placed close to the house without risk of flooding your basement, since the water is contained in the bed. This makes wicking beds a great alternative to swales on properties with sump pumps or basement water issues.
- No evaporation means no salting of soil. If you are watering your soils from the top with hard water, you risk accumulating salts, because the water evaporates and leaves the minerals behind. Eventually your soil will struggle to support plant life.
- They provide a lot of drainage in the event of a large downpour.
- Since they're raised, they will warm up quicker in the spring.
- You can easily attach cold frames to them.
- They are great for people with less mobility and strength as you don't have to haul heavy water containers.
- By using an intermediary tank, you can automate the watering process.

Disadvantages of Wicking Beds

Wicking beds do have some disadvantages as well:

- They cost more to install than in-ground swales and standard raised beds.
- They will freeze sooner in the fall than non-raised beds.
- There are additional freeze/thaw considerations that need to be taken into account, which is not required for conventional gardens.

Types of Wicking Beds

Reservoirs with Media

Most of the DIY sites for wicking beds focus on building beds that use media, a layer in between the soil and the water reservoir, as their wick. This is an easy and cheap way of supporting the soil on top of the reservoir. Gravel is the most common medium, but there are a number of materials that do the trick.

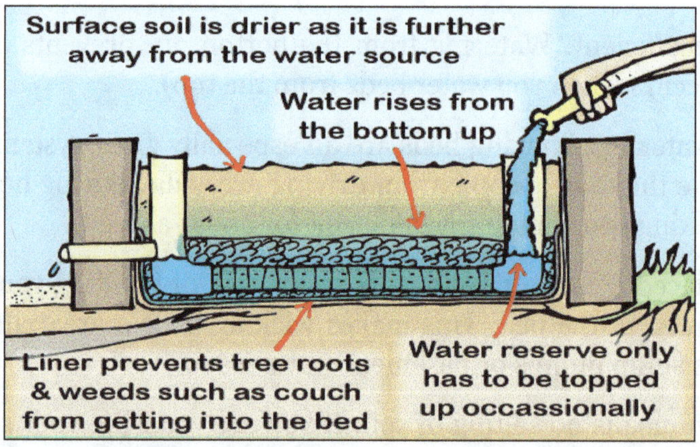

Media-less Beds

Beds without media require a false bottom that will allow the soil to be suspended above the water reservoir. Again, this wick system can be made from a variety of materials. Here are some examples of media-less wicking beds:

- Global Buckets
- Earth Box
- Phytopod

IMPACTS OF IRRIGATED AGRICULTURE

Value Chains and the Influence of Irrigation on Marketing and Processing

Most sub-Saharan Africa agricultural production is marketed and physically transformed prior to final consumption or industrial use. This includes a significant proportion of the foodstuffs produced on small-scale subsistence farms. Such farms frequently sell much of their output to meet urgent cash needs and then seek to buy staple foodstuffs later in the crop year. Given that a large proportion of total subSaharan Africa agricultural output is marketed, the efficiency of markets is vital both for the livelihoods of farm households and for other rural households that rely on the agriculture sector for employment.

Until the recent period of structural adjustment, grains, tropical beverages and most industrial crops grown in sub-Saharan Africa were processed and marketed either under systems operated by state-owned enterprises or under systems that involved some form of state price control. These have now largely been replaced by systems in which individuals, private firms and cooperatives undertake the processing and marketing. These systems rely on competitive forces to determine prices. The prices offered to producers of agricultural commodities depend on the costs incurred at each stage of the system and on the extent of competition at each point at which the commodity is traded. The former state-run systems were designed to be compatible with the production and market characteristics of the commodity in question. Under a free market, the structure of marketing systems evolves over time to suit the characteristics of the commodity, often through a painful process that involves the collapse of unsuitable systems and the failure of enterprises involved in them. As irrigation affects the production and market characteristics of agricultural commodities, it influences the evolution of marketing systems, their structure, and the efficiency at which they operate.

Spatial and Temporal Impacts Spatial Impacts

Irrigation affects the spatial distribution of agricultural production by allowing: (i) the growing of crops on land that was unable to sustain agriculture under rainfed conditions; (ii) the more intensive growing of existing crops; and (iii) the growing of alternative crops. For example, irrigation has allowed desert, semi-desert and low productivity rangeland in Namibia, Kenya and Sudan to be converted to the productionof fruits and cotton. In South Africa and other sub-Saharan Africa countries, irrigation has been utilized to raise the productivity of existing crop production, most notably the production of maize and vegetables. In central Kenya, irrigation has allowed some 5 000 ha that had been devoted to grazing and the growing of maize and other rainfed crops to be converted to rice production. In Swaziland, former low-veldt rangeland has been irrigated and utilized for the production of sugar cane. As irrigation normally leads to substantially higher yields, it has the effect of concentrating production spatially. This tendency is reinforced by engineering considerations relating to the supply of water, which frequently requires that irrigated land comprises a single contiguous area. The tendency for irrigation to lead to concentrated production applies to all irrigation regardless of the prior use of the land. The concentration of production raises the efficiency of marketing by allowing the exploitation of economies of scale, especially in transport. This applies both to road construction and maintenance and to vehicle use. Concentrated production also allows larger and more efficient processing units and reduces the distances over which the raw commodity is transported to a processing unit.

The benefits that stem from this differ between commodities depending on the value-to-weight ratio of the raw commodity, its perishability, the extent to which it is damaged during transport, and weight loss during processing. Of the main commodities grown under irrigation in sub-Saharan Africa, seed cotton is exceptionally bulky and sugar cane is both bulky and perishable. For both these commodities, but especially for sugar cane, the majority of the weight of the harvested commodity comprises low-value by-products that are removed during primary processing. For such commodities, the relative processing and marketing advantages afforded by irrigation are greater than for crops that are storable, are not easily damaged during transportation, are not exceptionally bulky, and which lose little weight during processing. The main sub-Saharan Africa crops that fall into this category are the staple grains, of which maize is the most important for food security. However, as yet, maize is largely insignificant as an irrigated crop. Vegetables and fruits vary significantly in terms of all the important variables that relate the spatial production characteristics of irrigation to processing and marketing efficiency. However, the generally labour-intensive nature of post-harvest activities for fruits and vegetables means that economies of scale in these activities are generally low. Most fruits and vegetables also lose little weight during "processing", which typically comprises only cleaning, grading and packing (fruits that are dried being the exception). Against this, most fruits and vegetables are highly perishable and easily damaged during handling and transport. On balance, the concentration of production afforded by irrigation probably gives fruits and vegetables less of a marketing and processing edge than it does for bulky crops such as cotton and sugar.

The concentrated production afforded by irrigation tends to lead to morecompetitive assembly markets as it increases the number of traders able to operate viably in a particular area. It also reduces the cost of tracing the origin of products and thereby increases the potential for small-scale farmers to sell to supermarkets and to other buying enterprises that seek to trace products to their source. Concentrated production also increases the feasibility of marketing by a single agency by reducing the costs of dealing with a set of small-scale farmers. The marketing agency could be either a farmers cooperative or association or an enterprise such as a ginnerycompany. Contract farming organized by ginnery companies and other processors has been particularly successful in sub-Saharan Africa as a replacement for former state-run single-channel marketing systems. The concentration afforded by irrigation helps increase the feasibility of contract farming by making it more difficult for farmers to engage in undetected side-selling. Against the generally positive impacts that spatially concentrated irrigated production has on marketing is the fact that the areas that can be irrigated most effectively may be distant from markets. For example, in Namibia, the only significant permanent rivers are in the extreme north and south of the country. Population density is low in the north and exceptionally low in the south. In the case of irrigation from the Orange River in the south, produce has to travel long distances to markets. Such new irrigation may only be commercially viable if new transport links are constructed, which may make the full investment package unattractive in terms of net benefits. While this can be a major problem that inhibits the development of irrigation, integrated irrigation development may have the desirable long-term side-effect of opening up new areas of the country to more widespread economic development. Indeed, governments may construct irrigation schemes in remote areas as part of a national strategy to decentralize development or stablilize rural communities. However, the general experience with the development of small-scale irrigation schemes in areas not effectively linked to markets has been disappointing.

Table: Irrigaed maize in sub-Sahran Africa for the baseline year.

Region	Irrigated maize area as % of total maize area	Irrigated maize production as % of total maize production
Central	0.00	0.00
Eastern	1.45	2.09
Gulf of Guinea	0.29	0.70
Indian Ocean Island	0.00	0.00
Republic of South Africa	3.47	8.99
Sothern	0.29	0.61
Sudano-Sahelian	9.11	8.86
Sub-Saharan Africa	1.36	3.40

Impacts on Quality

A switch from rainfed to irrigated production affects the quality characteristics of the commodity produced, including size, taste, smell, visual appearance, milling characteristics, and cooking properties. It also affects the extent to which these characteristics vary within a single harvested crop and between years. In so far as irrigation leads to healthier plants, the general size and quality of the produce is likely to be higher. However, it is possible that rapid growth may diminish the intensity of flavour and smell, reducing market value.

Irrigation typically has a major beneficial impact on the uniformity of the crop both between growers and over time. The main increase in uniformity is usually between years as under irrigation similar amounts of water can be applied on a timely basis each year. This contrasts with rainfed agriculture, where the timing and intensity of rainfall often varies markedly between years. Between farms, the introduction of irrigation into an area has the potential to lead to increased uniformity of the crop within that area to the extent that farmers under the former rainfed regime planted at different times or received different amounts of rainfall. However, this increased uniformity is dependent on equitable distribution of irrigation water between farmers, which is not always achieved. Another advantage stemming from increased planting-date uniformity under irrigation results from decreased accumulation of stage-dependent pests and pathogens. This is because populations of such pests and pathogens are not given the opportunity to expand by moving from plot to plot where excessively staggered planting provides ideal conditions for longer. Within an individual field, the contribution of irrigation to uniformity is likely to be minimal as all parts are likely to receive similar amounts of water under a rainfed regime. Indeed, it is possible for irrigation to lead to a less-uniform application of water within fields, as is often the case under furrow irrigation. In summary, the introduction of irrigation most commonly improves the overall level of quality and leads to less variation in quality between producers and between years. Reduced quality variation between producers serves to increase the efficiency of processing, especially where machinery is set for a specific standard of raw material, as is the case for most agricultural processing, particularly that involving milling. Reduced quality variation between years leads to two marketing benefits. First, it allows a set of irrigated farmers or an irrigated estate to develop a reputation for a particular quality of produce that attracts regular customers prepared to pay a premium price for dependable quality. Second, it assists producers to predict the quality of their crop. This helps them to sell forward with confidence and to lock into an assured producer price prior to harvest. presents an illustration of this impact.

Impacts on the Stability and Predictability of Production

Income stability is particularly important for small-scale producing households. This is because they typically lack the capacity to save and normally cannot borrow other than informally at very high rates of interest. This means that they are unable to moderate the impact of income instability on household expenditure. Consequently, they face severe hardship when income falls. The introduction of irrigation not only increases the level of crop output but also increases the stability of output from year to year. This tends to have a stabilizing impact on producer incomes, especially for internationally tradable commodities whose domestic prices are a function of international prices. For low-value perishable commodities, for which markets clear domestically, quantity and price movements tend to be offsetting. This moderates income instability in situations where common weather patterns affect the output of growers. Once irrigation is introduced for such commodities, output comprises stable irrigated production and unstable rainfed production. In this situation, the irrigated production reduces aggregate domestic instability in supply and prices. This in turn partially stabilizes the incomes of rainfed producers, provided demand for the commodity is price inelastic. The necessary degree of price inelasticity to lead to more stable gross incomes for rainfed producers increases as the share of irrigated output increases. Conversely, in such market circumstances, instability in rainfed production will always destabilize the gross incomes of irrigated producers unless irrigated output is for some reason unstable and correlated with rainfed output.

The generally stabilizing impact of the introduction of irrigation on farm household incomes stems from its effect on the stability of yield of a specific crop that was being grown previously under rainfed conditions. Should irrigation result in a switch of crops, it is conceivable that yields of the new irrigated crop may be less stable than the former rainfed crop. It is also possible that a switch to irrigated production may result in the growing of a crop for which prices tend to be less stable than those of the former crop. The stabilization of production has the important additional advantage of improving the accuracy with which producers can predict their output in advance of the production season. As with improvements to the consistency of quality, this increases their ability to sell forward and to enter into long-term contractual arrangements with input suppliers/processors. This in turn allows them to lock into a price in advance of the harvest and to eliminate price risk. In summary, the introduction of irrigation tends to reduce the risks facing producers by reducing the instability and improving the predictability of both yields and producer prices. As noted above, it also tends to improve and increase the predictability of quality. Together, these effects have the potential to improve the welfare of producers and to increase the net benefits from investing in irrigation.

Other Social Impacts

First, there is the income stability that better predictability brings as does the ability to diversify and thereby hedge against both market and climate shocks. This helps with household or group financial planning, and it also makes credit more manageable. It could also make credit more accessible, thereby facilitating further increases in production. Diversification also means that higher-value crops and crops with significant seasonal niche markets and added-value opportunities can often replace former subsistence systems, thereby allowing poor households better access to the local and national economies. Added-value opportunities often include grading, processing and packaging, many of which can be carried out within the farm or scheme boundary, thereby raising incomes significantly. As well as creating more on-farm jobs, the same measures increase the need for direct and indirect services concerned with the basic agricultural activity. Direct services might include cold-storage construction and operation, transportation, freight consolidation and the manufacturing and supply of packaging material and farm inputs. Indirect services are those associated with an economic growth point and are supplied by a broadening range of local artisanal skills, retail trade and equipment maintenance capacity. Finally, irrigation is beneficial also because it makes it easier to keep urban costs of basic commodities at affordable levels. However, when the supply of commodities increases, prices are likely to fall especially when shortfalls requiring imports are replaced by surpluses looking for a place in regional or global markets. Although it has been argued that adequately informed dynamic pricing capability is in place and sufficient to contain the more extreme results of price reductions, there will nonetheless be a detrimental effect on rainfed producers of the same commodities and the rural landless, further marginalizing them. This is especially the case the further such producers are from the employment opportunities represented by successful irrigation. Equally, ill-conceived or inflexible institutional measures have the potential to bankrupt farmer groups and their members, even where productivity is both efficient and high. This is especially risky where production becomes more specialized and dependent on inputs, monopolistic markets and rentseeking creditors and intermediaries. This situation is not helped by the fact that irrigated production is often subsidized (e.g. by low recurring-cost recovery) whereas rainfed farming seldom is.

IRRIGATION IN VITICULTURE

Irrigation in viticulture is the process of applying extra water in the cultivation of grapevines. It is considered both controversial and essential to wine production. In the physiology of the grapevine, the amount of available water affects photosynthesis and hence growth, as well as the development of grape berries. While climate and humidity play important roles, a typical grape vine needs 25-35 inches (635-890 millimeters) of water a year, occurring during the spring and summer months of the growing season, to avoid stress. A vine that does not receive the necessary amount of water will have its growth altered in a number of ways; some effects of water stress (particularly, smaller berry size and somewhat higher sugar content) are considered desirable by wine grape growers.

In many Old World wine regions, natural rainfall is considered the only source for water that will still allow the vineyard to maintain its *terroir* characteristics. The practice of irrigation is viewed by some critics as unduly manipulative with the potential for detrimental wine quality due to high yields that can be artificially increased with irrigation. It has been historically banned by the European Union's wine laws, though in recent years individual countries (such as Spain) have been loosening their regulations and France's wine governing body, the *Institut National des Appellations d'Origine* (INAO), has also been reviewing the issue.

In very dry climates that receive little rainfall, irrigation is considered essential to any viticultural prospects. Many New World wine regions such as Australia and California regularly practice irrigation in areas that couldn't otherwise support viticulture. Advances and research in these wine regions (as well as some Old World wine regions such as Israel), have shown that potential wine quality could increase in areas where irrigation is kept to a minimum and managed. The main principle behind this is controlled water stress, where the vine receives sufficient water during the budding and flowering period, but irrigation is then scaled back during the ripening period so that the vine then responds by funneling more of its limited resources into developing the grape clusters instead of excess foliage. If the vine receives too much water stress, then photosynthesis and other important processes such as nutrient storage could be impacted with the vine essentially shutting down. The availability of irrigation means that if drought conditions emerge, sufficient water can be provided for the plant so that the balance between water stress and development is kept to optimal levels.

Role of Water in Viticulture

The presence of water is essential for the survival of all plant life. In a grapevine, water acts as a universal solvent for many of the nutrients and minerals needed to carry out important physiological functions, and the vine receives these by absorbing the nutrient-containing water from the soil. In the absence of sufficient water in the soil, the root system of the vine may have difficulties absorbing these nutrients. Within the structure of the plant itself, water acts as a transport within the xylem to bring these nutrients to all ends of the plant. During the process of photosynthesis, water molecules combine with carbon derived from carbon dioxide to form glucose, which is the primary energy source of the vine, as well as oxygen as a by-product.

In addition to its use in photosynthesis, a vine's water supply is also depleted by the processes of evaporation and transpiration. In evaporation, heat (aided by wind and sunlight) causes water in the soil to evaporate and escape as vapor molecules. This process is inversely related to humidity with evaporation taking place at faster rates in areas with low relative humidity. In transpiration, this evaporation of water occurs directly in the vine, as water is released from the plant through the stomata that are located on the undersides of the leaves. This loss of water from the leaves is one of the driving factors that results in water being drawn up from the roots, and it also helps the vine combat against the effects of heat stress which can severely damage the physiological functions of the vine (somewhat similar to how perspiration works with humans and animals). The presence of adequate water in the vines can help keep the internal temperature of the leaf only a few degrees above the temperature of the surrounding air. However, if water is severely lacking then that internal temperature could jump nearly 18 °F (10 °C) warmer than the surrounding air which leads the vine to develop heat stress. The dual effects of evaporation and transpiration are called evapotranspiration. A typical vineyard in a hot, dry climate can lose as much as 1,700 U.S. gallons (6,400 l; 1,400 imp gal) of water per vine through evapotranspiration during the growing season.

Factors Influencing Irrigation

There are essentially two main types of irrigation; primary irrigation, which is needed for areas (such as very dry climates) that lack sufficient rainfall for viticulture to even exist, and supplemental irrigation where irrigation is used to "fill in the gaps" of natural rainfall to bring water levels to better numbers as well as to serve as a preventive measure in case of seasonal drought conditions. In both cases, both the climate and the vineyard soils of the region will play an instrumental role in irrigation's use and effectiveness.

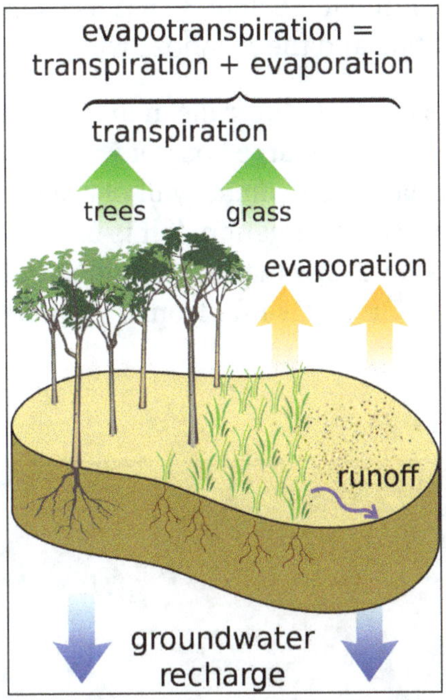

Climates with low humidity promote faster rates evapotranspiration which reduce the grapevine's water supply. These areas may need to utilize supplemental irrigation.

Impact of Different Climate Types

Viticulture is most commonly found in Mediterranean, continental and maritime climates with each unique climate providing its own challenges in providing sufficient water at critical times during the growing season. In Mediterranean climates irrigation is usually needed during the very dry periods of the summer ripening stages where drought can be a persistent threat. The level of humidity in a particular macroclimate will dictate exactly how much irrigation is needed with high levels of evapotranspiration more commonly occurring in Mediterranean climates that have low levels of humidity such as part of Chile and the Cape Province of South Africa. In these low humidity regions, primary irrigation may be needed, but in many Mediterranean climates the irrigation is usually supplemental. The amount of precipitation that occurs during spring and summer months is also important. For example, Tuscany receives an average of 8 inches (200 mm) of rainfall during the months of April through June - the period that includes flowering and fruit set, when the water is most crucial. While fluctuations in rainfall do occur, the amount of natural precipitation, combined with water holding capacity of soil, is typically sufficient to result in healthy harvest. In contrast, Napa Valley only gets 2.4 inches (60 mm), on average, during the same period of time, often in an erratic pattern (some years seeing more, some years seeing only trace amounts of rain), and most appellations in Central and Southern California (both along the coast and inland) receive even less than that, necessitating supplemental irrigation.

Continental climates are usually seen in areas further inland from the coastal influences of oceans and large bodies of water. The difference from the average mean temperature of its coldest and hottest months can be quite significant with moderate precipitation that usually occurs in the winter and early spring. Depending on the water retaining ability of the soil the grapevine may receive enough water during this period to last throughout the growing season with little if any irrigation needed. For soils with poor water retention, the dry summer months may require some supplemental irrigation. Examples of continental climates that use supplemental irrigation include the Columbia Valley of Washington State and the Mendoza wine region of Argentina.

Maritime climates tend to fall between Mediterranean and continental climates with a moderate climate that is tempered by the effects of a large body of water nearby. As with Mediterranean climates, the humidity of the particular macroclimate will play a significant role in determining how much irrigation is needed. In most cases irrigation, if it is used at all, will only be supplemental in years where drought may be an issue. Many maritime regions, such as Rias Baixas in Galicia, Bordeaux and the Willamette Valley in Oregon, suffer from the diametric problem of having too much rain during the growing seasons.

Impact of Different Soil Types

Soil can have a significant impact on the potential quality of wine. While geologist and viticulturist are not exactly sure what type of immutable or *terroir* based qualities that soil can impart on wine, there is near universal agreement that a soil's water retention and drainage abilities play a primary role. Water retention refers to the soil's ability to hold water. The term "field capacity" is used to describe the maximum amount of water that deeply moistened soil will retain after normal drainage. Drainage is the ability of water to move freely throughout the soil. The ideal circumstance is soil that can retain sufficient amount of water for the grapevine but drains well enough to where the soil doesn't become water-logged. Soil that doesn't retain water well encourages the vine to easily sleep into water stress while soil that doesn't drain well runs of the risk of water-logged roots being attacked by microbial agents that consume all the soil nutrients and end up starving the vine.

The depth, texture and composition of soils can influence its water retaining and draining ability. Soils containing large amounts of organic material tend to have the highest water retention abilities. These types of soils include deep loams, silty soils like what is typically found on the fertile valley floors such as in the California's Napa Valley. Clay particles have the potential to remain in colloidal suspension for long periods of time when it is dissolved in water. This gives clay-based soils the potential to retain significant amount of water such as the clay soils of the Right bank Bordeaux region of Pomerol. Many regions with these types of water retaining soils have little need for irrigation, or if they do it is usually supplemental during periods of drought. Soils with poor water retention include sand and alluvial gravel based soils such as those found in the Barolo and Barbaresco zones of Italy or in many areas of South Australia. Depending on the climate and amount of natural rainfall, areas with poor water retention may need irrigation.

Just as having too little water is detrimental to the grapevine, so too is having too much. When vines become water-logged they become a target for various microbial agents such as bacteria and fungi that compete with the vine for nutrients in the soil. Additionally excessively moist soil is poor conductor of valuable heat radiating from the ground. In general wet soils are cold soils which can be especially problematic during the flowering causing poor berry set that could lead to coulure. It also becomes an issue during the ripening stage when vines in cool-climate regions may need additional heat radiated from the ground in order to sufficiently ripen its fruit (an example of this is the slate-based vineyards of the Mosel in Germany). Therefore, well draining soils are considered very conducive to producing quality wine. In general light-textured (such as sand and gravel) and stony soils tend to drain well. Soils heavy soils and those with high proportions of organic matter also have the potential to drain well if they having a crumbling texture and structure. This texture relates to the friability of the soil which can come from earthworms and other organisms that have burrowed tunnels throughout the soil. Much like rocks, these tunnels give water a freer passageway through soil and contributes to its drainage.

Measuring Soil Moisture

Because of the problems associated with water-logged and wet soils, it is important for viticulturist to know how much water is currently in the soil before deciding if and how much to irrigate. There are several methods of evaluating soil moisture. The most basic is simple observation and feeling of the soil, however this has its limitations since the subsoil may be moist while the surface soil appears dry. More specific measurements can be attained by using tensiometers which evaluates surface tension of water extracted from the soil. The presence of water in the soil can be measured by neutron moisture meters that utilize an aluminium tube with an internal neutron source that detect the subtle change between the water in the soil. Similarly, gypsum block placed throughout the vineyard contain an electrode that can be used to detect the electrical resistance that occurs as the soil dries and water is released by evaporation. Since the 1990s there has been greater research into tools utilizing time-domain reflectometry and capacitance probes. In addition to monitoring for excessive moisture, viticulturists also keep an eye for signs of water stress due to severe lack of water.

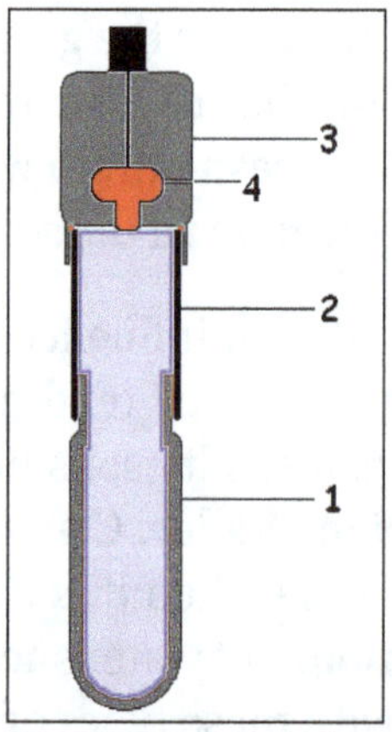

Tensiometers can be used to measure soil moisture. The components of this example include (1) porous cup, (2) water-filled tube (3) sensor-head and a (4) pressure sensor.

Irrigation Systems

There are several methods of irrigation that can be used in viticulture depending on the amount of control and water management desired. Historically, surface irrigation was the most common means using the gravity of a slope to release a flood of water across the vineyard. In the early history of the Chilean wine industry, flood irrigation was widely practiced in the vineyards using melted snow from the Andes Mountains channeled down to the valleys below. This method provided very little control and often had the adverse effect of over-watering the vine. An adaption of method was the furrow irrigation system used in Argentina where small channels ran through the vineyard providing irrigation. This provide slightly more control since the initial amount of water entering the channels could be regulated, however the amount that each vine received was sporadic.

Sprinkler irrigation involves the installation of a series of sprinkler units throughout the vineyard, often spaced as several rows about 65 feet (20 m) apart. The sprinklers can be set on an electronic timer and release predetermined amount of water for a set period of time. While this provides more control and uses less water than flood irrigation, like furrow irrigation the amount that each individual wine receives can be sporadic. The irrigation system that provides the most control over water management, though conversely the most expensive to install, is drip irrigation. This system involved long plastic water supply lines that run down each row of vines in the vineyard with each individual grape vine having its own individual dripper. With this system, a viticulturist can control the precise amount of water that each grapevine gets down to the drop. An adaption of this system, potentially useful in areas where irrigation may be banned, is underground subirrigation where precise measurements of water is delivered directly to the root system.

When and how Much?

With abundant water, a grapevine will produce shallow root systems and vigorous growths of new plant shoots. This can contribute to a large, leafy canopy and high yields of large grape berry clusters that may not be sufficiently or physiologically ripe. With insufficient water, many of the vine's important physiological structures, including photosynthesis that contributes to the development of sugars and phenolic compounds in the grape, can shut down. The key to irrigation is to provide just enough water for the plant to continuing function without encouraging vigorous growth of new shoots and shallow roots. The exact amount of water will depend on a variety of factors including how much natural rainfall can be expected as well as the water retaining and drainage properties of the soil.

Water is very crucial during the early budding and flowering stages of the growing season. In areas where there is not sufficient rainfall, irrigation may be needed during this time in the spring. After fruit set, the water needs for the vine drop and irrigation is often withheld till the period of *veraison* when the grapes begin to change color. This period of "water stress" encourages the vine to concentrate its limited resources into lower yields of smaller berries creating a favorable skin to juice ratio that is often desirable in quality wine production. The benefits or disadvantages of irrigation during the ripening period itself is a matter of debate and continuing research in the wine growing community. The only area of mostly agreement is the disadvantages of water close to harvest after a prolonged dry period. Grapevines that have been subjected to prolonged water stress have a tendency to rapidly absorb large amounts of water if its provided. This will dramatically swell the berries, potentially causing to them crack or burst which will make the prone to various grape diseases. Even if the berries do not crack or burst, the rapid swelling of water will cause a reduce concentration in sugars and phenolic compounds in the grape producing wines with diluted flavors and aromas.

Water Stress

One of the goals of controlled, mild water stress is to discourage the formation of excess new plant growths (a bud pictured) which will compete with the developing grape clusters for the vine's limited resources.

The term water stress describes the physiological states that grapevines experience when they are deprived of water. When a grapevine goes into water stress one of its first functions is to reduce the growth of new plant shoots which compete with the grape clusters for nutrients and resources. The lack of water also keeps the individual grape berries down to a smaller size which increase its skin to juice ratio. As the skin is filled with color phenols, tannin and aroma compounds, the increase in skin-to-juice ratio is desirable for the potential added complexity the wine may have. While there is disagreement over exactly how much water stress is beneficial in development grapes for quality wine production, most viticulturist agree that some water stress can be beneficial. The grapevines in many Mediterranean climates such as Tuscany in Italy and the Rhone Valley in France experience natural water stress due to the reduced rainfall that occurs during the summer growing season.

At the far extreme is severe water stress which can have detrimental effects on both the vine and on potential wine quality. To conserve water, a vine will try to conserve water by limiting its loss through transpiration. The plant hormone abscisic acid triggers the stomata on the underside of the plant leaf to stay close in order to reduce the amount of water that is evaporated. While conserving water this also has the consequences of limiting the intake of carbon dioxide needed to sustain photosynthesis. If the vine is continually stressed it will keeps it stomata closed for longer and longer periods of time which can eventually cause photosynthesis to stop all together. When a vine has been so deprived of water it can exceed what is known as its permanent wilting point. At this point, the vine can become permanently damaged beyond recovery even if later watered. Viticulturists will carefully watch the plant for signs of severe water stress. Some of the symptoms include:

- Flaccid and wilting tendrils.
- (During Flowering) Flower clusters that are dried out.
- Wilting of young grape leaves followed by maturer leaves.
- Chlorosis signaling that photosynthesis has stopped.
- Necrosis of dying leaf tissue which leads to premature leaf fall.
- Finally, the grape berries themselves start to shrivel and fall off the vine.

The effectiveness of water stress is an area of continuing research in viticulture. Of particular focus is the connection between yield size and the potential benefits of water stress. Since the act of stressing the vine does contribute to reduce photosynthesis-and by extension, reduce ripening since the sugars produced by photosynthesis is needed for grape development-it is possible that a stressed vine with high yields will only produce lots of under ripe grapes. Another interest of study is the potential impact on white grape varieties with enologists and viticulturists such as Cornelius Van Leeuwen and Catherine Peyrot Des Gachons contending that white grape varieties lose some of their aromatic qualities when subjugated to even mild forms of water stress.

Partial Rootzone Drying

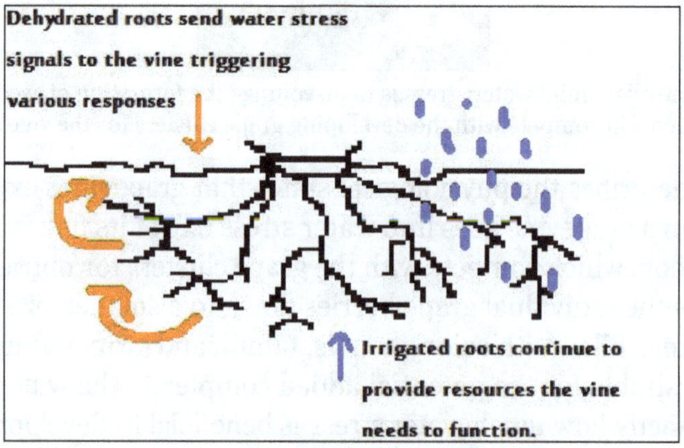

One irrigation technique known as partial rootzone drying (or PRD) involves "tricking" the grapevine into thinking it is undergoing water stress when it is actually receiving sufficient water supply. This is accomplished by alternating drip irrigation to where only one side of the grapevine receives water at a time. The roots on the dry side of the vine produce abscisic acid that triggers some of the vine's physiological responses to water stress - reduced shoot growth, smaller berries size, etc. But because the vine is still receiving water on the other side the stress doesn't become so severe to where vital functions such as photosynthesis is compromised. Partial rootzone drying has been shown to significantly increase a vine's water use efficiency. While PRD is shown to slightly reduce leaf area, this is generally not a problem as overall yield is unaffected.

Partial Rootzone Drying

One irrigation technique known as partial rootzone drying (or PRD) involves "tricking" the grapevine into thinking it is undergoing water stress when it is actually receiving sufficient water supply. This is accomplished by alternating drip irrigation to where only one side of the grapevine receives water at a time. The roots on the dry side of the vine produce abscisic acid that triggers some of the vine's physiological responses to water stress - reduced shoot growth, smaller berries size, etc. But because the vine is still receiving water on the other side the stress doesn't become so severe to where vital functions such as photosynthesis is compromised. Partial rootzone drying has been shown to significantly increase a vine's water use efficiency. While PRD is shown to slightly reduce leaf area, this is generally not a problem as overall yield is unaffected.

Criticism and Environmental Issues

The practice of irrigation has its share of criticism and environmental concerns. In many European wine regions the practice is banned under the belief that irrigation can be detrimental to quality wine production. However, in the early 21st century some European countries have relaxed their irrigation laws or reevaluated the issue. Of the criticisms leveled towards irrigation, the most common is that it disrupts the natural expression of *terroir* in the land as well as the unique characteristics that comes with vintage variation. In regions that do not practice irrigation, the quality and styles of wines can be dramatically different from vintage to vintage depending on weather conditions and rainfall. Irrigation's contribution to the broader globalization of wine is criticized as promoting a homogenization or "standardization" of wine.

Other criticisms center around the broader environmental impact of irrigation on both the ecosystem around the vineyard as well as the added strain on global water resources. While advances in drip irrigation has reduced the amount of waste water produced by irrigation, the irrigation of substantial tracts of land in areas like the San Joaquin Valley in California and the Murray-Darling Basin of Australia requires massive amounts of water from dwindling supplies. In Australia, the centuries-old practice of flood irrigation used in places like the Murrumbidgee Irrigation Area caused severe environmental damages from water-logging, increase salination and raising the water tables. In 2000, the Australian government invested over A$3.6 million into research on how to minimize the damage caused by extensive irrigation. In 2007, concerns about ecological damage to the Russian River caused government officials in California to take similar measures to cut back water supplies and promote more efficient irrigation practices.

Other uses for Irrigation Systems

In addition to providing water for plant growth and development, irrigation systems can also be used for alternative purposes. One of the most common is the dual application of fertilizer with water in a process known as fertigation. Commonly used in drip irrigation systems, this method allows similarly regulate control over how precisely how much fertilizer and nutrients that each vine receives. Another alternative use for sprinkler irrigation systems can occur during the threat of winter or spring time frost. When temperature drop below 32 °F (0 °C), the vine is at risk of developing frost damage that could not only ruin the upcoming years harvest but also kill the vine. One preventive measure against frost damage is to use the sprinkler irrigation system to coat the vines with a protective layer of water that freezes into ice. This layer of ice serves as insulation keeping the internal temperature of the vine from dropping below the freezing mark.

References

- Delefosse, Matthieu (2015). "Seeing The Unseen—Bioturbation In 4D: Tracing Bioirrigation In Marine Sediment Using Positron Emission Tomography And Computed Tomography". PLOS ONE. 10 (4): e0122201. Doi:10.1371/journal.pone.0122201. PMC 4383581
- About-irrigation, irrigation, soil-and-water, farm-management, agriculture: vic.gov.au, Retrieved 1 March, 2019
- J. Robinson (ed) "The Oxford Companion to Wine" Third Edition pp. 236-263, 361-362, 507-566, 635-641, 709-763 Oxford University Press 2006 ISBN 0-19-860990-6
- From-the-bottom-up-a-diy-guide-to-wicking-beds: permaculturenews.org, Retrieved 2 April, 2019
- Elhindi, Khalid, El-Hendawy, Salah, Abdel-Salam, Eslam, Elgorban, Abdallah, & Ahmed, Mukhtar. (2016). Impacts of fertigation via surface and subsurface drip irrigation on growth rate, yield and flower quality of Zinnia elegans. Bragantia, 75(1), 96-107. Epub December 22, 2015.https://dx.doi.org/10.1590/1678-4499.176

2

Types and Methods of Irrigation

Several types of methods are used in irrigation such as check basin method, furrow irrigation, sprinkler irrigation, drip irrigation, pot irrigation, tidal irrigation, surface irrigation, micro-irrigation, sprinkler irrigation etc. The topics elaborated in this chapter will help in gaining a better perspective about these traditional and modern methods of irrigation.

TRADITIONAL IRRIGATION METHODS

Traditional irrigation Methods are the methods of irrigation that were used in the earlier years. Even today some small farms in rural areas adopt these. Although they are cheaper than the modern methods, they are not nearly as efficient. They require human or animal labour to function.

Check Basin Method

In this method, the whole field is divided into basins according to the capacity of water. Basins are connected through a 'Dhora' (A small drain type flow way), which has raised earthen walls on both sides. 'Dhora' is of two types, one is the main 'Dhora' and the other 'Dhora' is connected to basins. Size of basins are made according to the inflow of water.

These basins are surrounded by small furrows. Branch 'Dhora' flows towards the slope from the main 'Dhora'. If the slope of branch 'Dhora' is steep, 'mooonja' or polythene is spread in it to prevent erosion of sides. The main source of water is located at the highest place in the field.

The width of drains is affected by factors like flow of water, percentage, slope and structure of the ground etc. The length of 'Dhora'is different depending on the basis of slope and formation of the fields. This method is also prevalent in India as it does not cause any burden on the farmer.

This method has the following advantages:

1. It is the best method of irrigation for leveled fields.
2. It does not require any technical knowledge.
3. This method is more useful in soils having lesser infiltration.
4. In this method, rain water stays in basins, hence soil erosion is not caused.

5. It has lesser economic investment.
6. It irrigates more area.
7. Crops gets sufficient water.

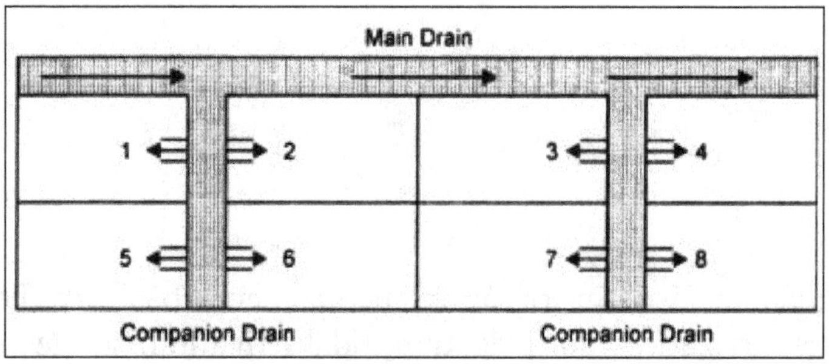

Check Basin Method.

Following are disadvantages of check basin method:

1. Due to seepage in drains, wastage of water is caused.
2. Machines cannot be used m this method because during spray of insecticides or fertilizers, the earthen walls of basins are damaged.
3. There is imbalance in distribution of labour. After growth of crops, water reaches the basins in disproportionate quantity thereby causing wastage of water.
4. Creation of problem of water logging.

Furrow Irrigation Method

Furrow irrigation method is resorted to where crops are one grown in rows. Along the side of rows of crops, 'Dol'is formed, and in between two such 'Dols', a furrow is formed in which water flows for irrigation. The quantity of flow of water depends on demand of water by plants and the rate of infiltration.

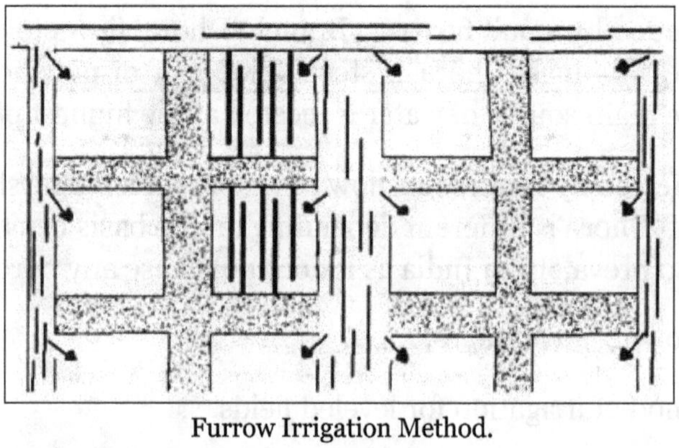

Furrow Irrigation Method.

In different situations, different furrow methods are used. They are mainly of five types:

1. Slopy Furrow.
2. Leveled Furrow.

3. Contour Furrow.
4. Serial Furrow.
5. Corrugated Furrow.

Advantages

1. Large areas can be irrigated at a time.
2. It saves labour since once the furrow is filled, it is not necessary to give water a second time.
3. It is a comparatively cheaper method.
4. Plants gets suitable quantity of water by this method.

Disadvantages

1. Due to imbalance in flow of water, wastage of water is caused in it.
2. It is not suitable in all types of crops.
3. Making 'Dol' for drains requires more labour information.
4. Due to filling of excess water, there is risk of underground salts coming up to the surface layer.

Strip Irrigation Method

In strip irrigation method, fields are divided into strips of different size. A boundary called 'Med' is formed to separate the strips. These strips are constructed according to the slope. The source of water is situated at the highest place in the field from where the whole field can get the flow of water.

The width of strips is decided as per quantity of water. More wastage of water is caused if strips are wider. Length of strip is decided by the slope of land and its structure. Effect of soil composition is also visible on it.

Advantages

1. It is possible to irrigate more area at a lesser expenditure.
2. It requires less labour.
3. Method of irrigation is easy and it causes lesser erosion.

Disadvantages

1. It is not suitable for all types of crops.
2. It is not possible to get balanced supply of water.
3. It is not suitable for all soil compositions.

Basin Irrigation Method

This irrigation method is more suited for horticulture development. In this method, a raised platform called 'Thanvla' is formed around trees or bushes and they are connected with each other through drains and the water reaches from one tree to the other. This method is not suitable for crops.

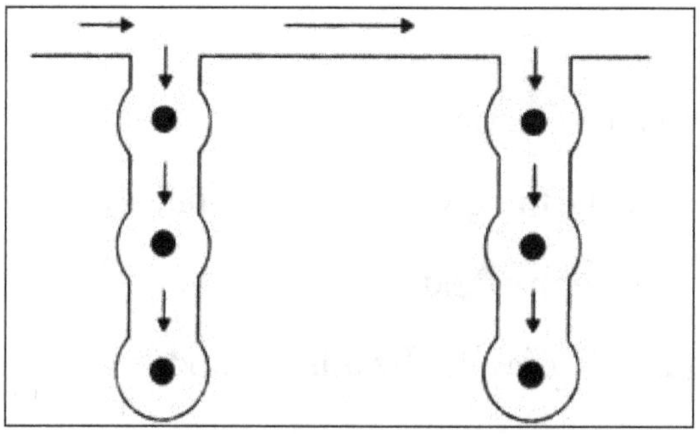

Basin Irrigation Method.

Advantages

1. It saves time. Once the water is opened, it reaches other trees automatically.
2. Its economic investment is less.
3. It is beneficial for more trees.

Disadvantages

1. It is not useful for all crops.
2. Wastage of water is caused in it.
3. Diseases spread in trees.

MANUAL IRRIGATION METHODS

Manual irrigation systems are very simple, but effective methods for making water available to crops. Manual irrigation systems are easy to handle and there is no need for technical equipment. But it is important that they are constructed correctly to avoid water loss and crop shortfall. The systems allow for high self-help compatibility and have low initial capital costs. They can be used in almost every area, but they are especially adapted for arid areas where evaporation rates are high.

Manual irrigation systems are easy to handle, require no technical equipment and are therefore generally cheap (in contrast to high-tech systems such as sprinkler irrigation or subsurface drip irrigation). But they need high labour inputs. A common and very simple technique for manual irrigation is for instance the use of watering cans as it can be found in peri-urban agriculture around large cities in some African countries. A more sophisticated and very water-efficient type of manual irrigation system is small-scale drip irrigation with buckets. Beside these systems, there are many other methods for manual irrigation, which are easy to install and simple to use. In general, all of these methods have high self-help compatibility and a relatively high performance. Therefore such systems are also called HELPFUL irrigation methods: High-frequency, Efficient, Low-volume, Partial-area, Farm-Unit, and Low-cost.

Low-cost Drip Irrigation System

A low-cost "farm-kit system" with a 1000 litres water tank can service up one-eight of an acre.

Simple drip irrigation (in contrary to high tech drip irrigation systems) uses low-cost plastic pipes cut to the appropriate lengths laid on the ground to irrigate vegetables, field crops and orchards. Small holes in the hose allow water to drip out and keep the base of the plant wet without wasting any water.

Watering Cans

Irrigation by watering cans is a very basic way but is still widely used. This creates a lot of work for the labours especially if this technique is used for large fields. A common way to make this work easier is a carry-pole across the shoulders. The field worker is able to carry two big watering cans in each side and the irrigation water can de distributed equally on the field. A rose can be added to the watering-can to create a "sprinkler effect".

With watering cans, the field worker is able to irrigate very specific and only where it is necessary.

Farmers in Africa irrigate a field with watering cans. The rose on the top of the outlet creates a sprinkler effect. A carry-pole across the shoulders as done in many parts of Asia would simplify the irrigation work.

Pitcher Irrigation

A very basic subsurface method consists in placing porous clay jars (or pots) in shallow pits dug for this purpose. Soil is then packed around the necks of the jars so that their rims protrude a few centimetres above the ground surface. Water is poured into the jars either by hand or by means of a flexible hose connected to a water source. Since the walls of the pots are porous (make sure to use unglazed pots), the water can seep slowly out and reach the roots of the plants. The jars can be made of locally available clay: they are of no standard shape, size, wall thickness or porosity. Instead of a clay or earthenware pod, also the sweet monkey orange fruit (Strychnos spinosa) can be used when it has been dried and the top cut off. The jar should be filled up regularly (especially in arid areas) and has to be changed if there are big cracks that the water percolates without reaching the roots.

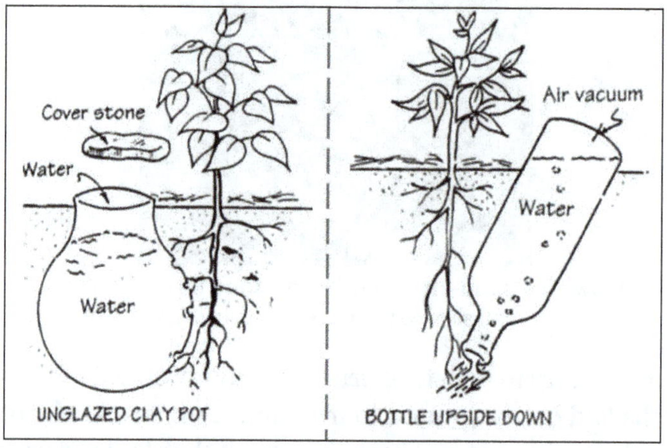

Left: Clay pot irrigation method. Right: Bottled irrigation method is also effective and simple.

Bottle Irrigation

Pitcher irrigation can also be done using a bottle. The bottle is first filled and than placed with its neck into the soil next to a plant, so it stands upside down. The dense soil hinders the water from leaving the bottle immediately. Instead, it gets released slowly and directly besides the roots, so it is available to the plant for a longer time and the water cannot evaporate directly.

Porous and Sectioned Pipes

Another variation of pitcher irrigation uses porous pipes instead of pots to spread water along a continuous horizontal band in the soil, rather than at discrete locations. As such, the porous pipe method is more suitable for closely spaced row crops grown in beds, such as vegetable crops. One end of the porous pipe is made protrude above ground that the operator is able to refill it with water as soon as it is necessary. In contrast to subsurface drip irrigation, the porous pipe distributes the water over the whole length and not only where it is perforated. That means it is less effective and water loss is slightly higher.

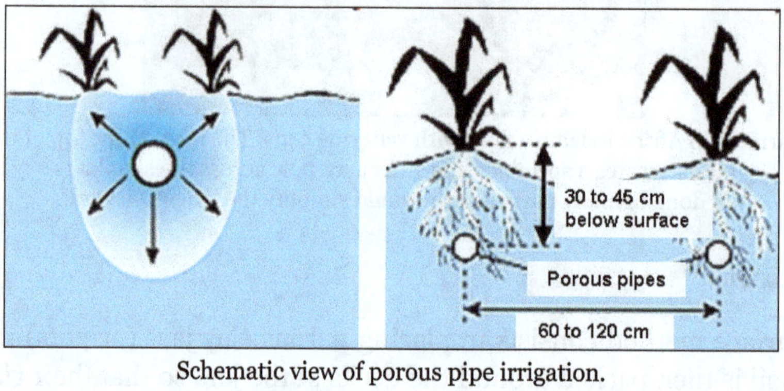

Schematic view of porous pipe irrigation.

Perforated Plastic Sleeves

An interesting variation of the pitcher method is the use of thin plastic sheeting to form a sleeve-like casing. To define its comparative usefulness better, the method should be tested side by side with alternative methods of irrigation. To date, this has not been done systematically.

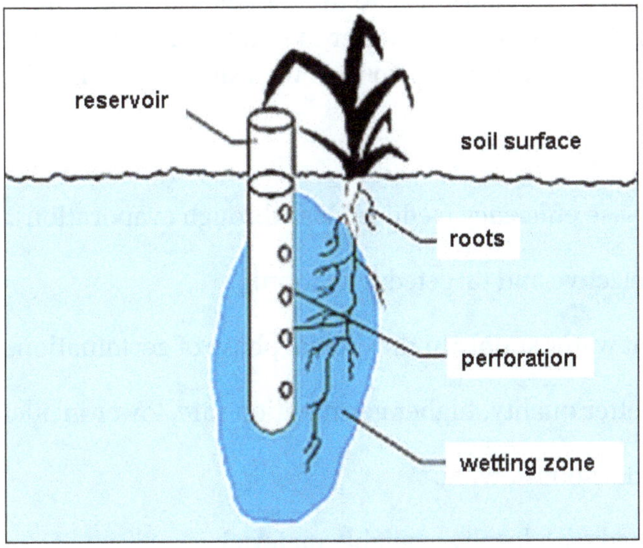

The plastic sleeve method is not tested systematically and therefore it is difficult to estimate its performance.

Costs Considerations

All of the described systems are very cheap. Many of them can be made out easily available material (e.g. old buckets or bottles) or renewable resources (bamboo). This reduces the costs significantly. However, high labour inputs are required for operation and maintenance.

Operation and Maintenance

As water is brought into the system manually, this requires high labour input. Moreover, it is important to check the systems regularly to prevent blockages and leakages. If there are any problems it should be cleaned and fixed as fast as possible to prevent damages on crops. Furthermore, there are several techniques to improve the production and avoid water loss on the fields.

Health Aspects

If wastewater is used for the manual irrigation process, there are potential health risks if water is not properly pre-treated (i.e. inadequate pathogen reduction). If poorly treated wastewater is applied. Appropriate pre-treatment should precede any irrigation scheme to limit health risks to those who come in contact with the water. As well, depending on the degree of treatment that the effluent has undergone, it may be contaminated with the different chemicals that are discharged into the system. When effluent is used for irrigation, households and industries connected to the system should be made aware of the products that are and are not appropriate for discharging into the system. Drip irrigation and subsurface drip irrigation are the only types of irrigation that should be used with edible crops, and even then, care should be taken to prevent workers and harvested crops from coming in contact with the treated effluent. Despite safety concerns, irrigation with effluent is an effective way to recycle nutrients and water.

Applicability

Manual irrigation methods are appropriate for small-scale farming or backyard gardening irrigation in dry and arid climates where water is scarce. All the different designs reduce water evaporation. It allows people to grow its own food with simple but effective techniques.

Advantages

- Improved water-use efficiency (reduced loss through evaporation).
- Well directed, selective and targeted irrigation.
- Ensures constant water supply in the crucial phase of germination.
- Higher yields, better quality, higher germination rate, lower incidence of pest attack.
- Facilitates pre-monsoon sowing.
- Can be constructed with locally available material.
- Low investment costs.

Disadvantages

- Labour intensive.
- User need a basic training to install and use the correct most of the method.
- If the water is not properly filtered and the equipment not properly maintained, it can result in clogging.
- Manual subsurface drip irrigation avoids the high capillary potential of traditional surface applied irrigation, which can draw salt deposits up from deposits below.

MODERN IRRIGATION METHODS

Modern methods of Irrigation are more efficient systems of irrigation that were invented in the recent decades. These help us use water economically without wastage.

Pot Irrigation Method

Pot irrigation method is more suitable for areas having scanty rainfall. In saline areas where flow irrigation is not suited, pot irrigation method is successful. An earthen pitcher is used in this method. The pitcher is fixed in the ground up to neck.

Holes are made in the pitcher and water is filled in it so that seepage of water through the holes keeps the nearby soil moist. Water is filled in these pitchers at regular intervals. This method can be considered as an alternative of drip irrigation method.

Pot irrigation method can be adopted in the following conditions:

1. Unlevelled land which is uneven.
2. Area having maximum shortage of water.
3. Such difficult areas where supply of fruits and vegetables is difficult and they are costly.
4. Where there is saline water, making surface irrigation difficult.

In this method, distribution of humidity around sides of pitcher is affected by many factors, mainly size of the pitcher, seepage of water per unit of area and type of soil. Humidity is spread in the same proportion as the size of the pitcher. Distance of pitcher also affects the moist area. Normally, distance between two pitchers should be kept so much that the humid area between them does not overlap.

Advantages

1. In this method, only the area near the pot gets irrigated and not the whole area.
2. Evaporation of water is minimum in this method.
3. Water seepage below the ground is also in minimum quantity.
4. It is the best method for horticulture crops and vegetables.
5. Once the pitchers are fixed, irrigation can be done for six years, which reduces expenditure.
6. It needs minimum technical knowledge.

Defects

1. Irrigation in this method is possible in a limited area.
2. This method requires clean water because unclean water would cause blockage of minor holes, which would not be able to provide moisture any longer.
3. It is costly to draw out pitchers again and again and re-fix them.
4. It is not suitable for every crop.

CENTER PIVOT IRRIGATION

Center-pivot irrigation (sometimes called central pivot irrigation), also called water-wheel and circle irrigation, is a method of crop irrigation in which equipment rotates around a pivot and crops are watered with sprinklers. A circular area centered on the pivot is irrigated, often creating a circular pattern in crops when viewed from above (sometimes referred to as *crop circles*). Most center pivots were initially water-powered, and today most are propelled by electric motors.

A satellite image of circular fields characteristic of center pivot irrigation.

Linear/Lateral Move Irrigation Machines

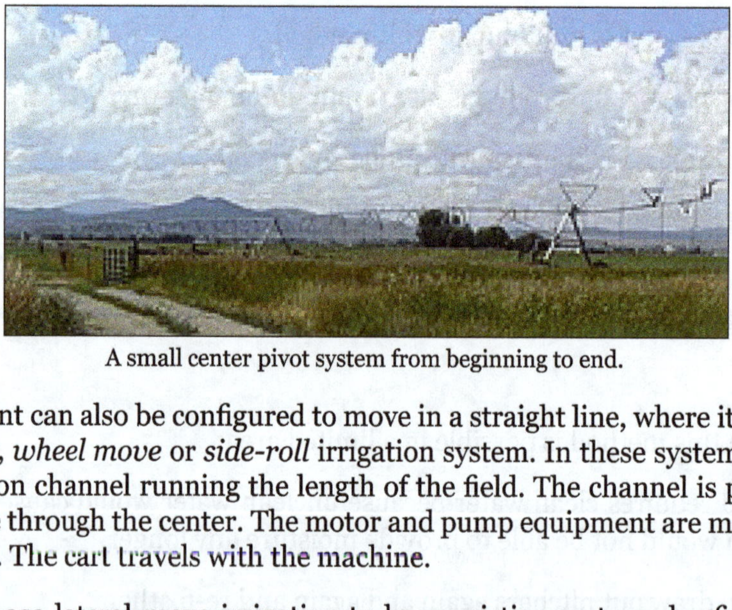

A small center pivot system from beginning to end.

Irrigation equipment can also be configured to move in a straight line, where it is termed a *lateral move, linear move, wheel move* or *side-roll* irrigation system. In these systems the water is supplied by an irrigation channel running the length of the field. The channel is positioned either at one side or in a line through the center. The motor and pump equipment are mounted on a cart by the supply channel. The cart travels with the machine.

Farmers might choose lateral-move irrigation to keep existing rectangular fields. This can help them convert from furrow irrigation. Lateral-move irrigation is far less common, relies on more complex guidance systems, and requires additional management compared to center pivot irrigation. Lateral-move irrigation is common in Australia. There, systems are usually between 500 and 1,000 meters long.

Benefits

Center-pivot irrigation uses less labor than many other surface irrigation methods, such as furrow irrigation. It also has lower labor costs than ground-irrigation techniques that require digging of channels. Also, center-pivot irrigation can reduce the amount of soil tillage. Therefore, it helps reduce water runoff and soil erosion that can occur with ground irrigation. Less tillage also encourages more organic materials and crop residue to decompose back into the soil. It also reduces soil compaction.

In the United States early settlers of the semiarid High Plains were plagued by crop failures due to cycles of drought, culminating in the disastrous Dust Bowl of the 1930s. Only after World War II when center pivot irrigation became available did the land mass of the High Plains aquifer system transform into one of the most agriculturally productive regions in the world.

Negative Effects

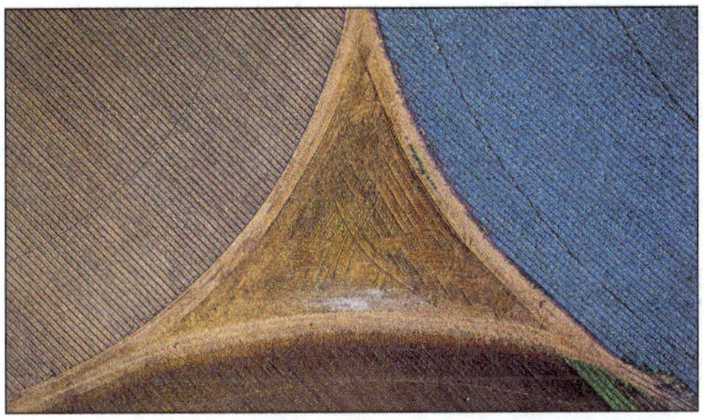

Fossil water is a non-renewable resource. Groundwater levels decrease when the rate of extraction by irrigation exceeds the rate of recharge. By 2013 it was shown that as the water consumption efficiency of center-pivot irrigation improved over the years, farmers planted more intensively, irrigated more land, and grew thirstier crops.

In parts of the United States, sixty years of the profitable business of intensive farming using huge center-pivot irrigators has emptied parts of the Ogallala Aquifer (also known as the High Plains Aquifer). One of the world's largest aquifers, it covers an area of approximately 174,000 mi^2 (450,000 km^2) in portions of the eight states of South Dakota, Nebraska, Wyoming, Colorado, Kansas, Oklahoma, New Mexico, and Texas, beneath the Great Plains in the United States.

In 1950, irrigated cropland covered 250,000 acres. With the use of center-pivot irrigation, nearly three million acres of land were irrigated in Kansas alone. At some places, during maximum extraction, the water table dropped more than five feet (1.5 m) per year. In extreme cases, wells had to be greatly deepened to reach the steadily falling water table. In some places in the Texas Panhandle, the water table has been drained (dewatered). "Vast stretches of Texas farmland lying over the aquifer no longer support irrigation. In west-central Kansas, up to a fifth of the irrigated farmland along a 100-mile (160 km) swath of the aquifer has already gone dry." It would take hundreds to thousands of years of rainfall to replace the groundwater in the dried up aquifer.

Role in Culture

In the 21st century, recognition of the significance of the High Plains Aquifer has led to increased coverage from regional and international journalists. A May 2013 *New York Times* article recounts the relentless decline of parts of the High Plains Aquifer System.

Writer Emily Woodson characterized the increased use of the center pivot irrigation system as part of a profound attitude shift towards modernism (expensive tractors, center-pivot irrigation, dangerous new pesticides) and away from traditional farming that took place in the mid-1970s and 1980s in the United States. A new generation chose high-risk, high-reward crops such as irrigated corn or peanuts, which require large quantities of ground water, fertilizer and chemicals. The new family farm corporations turned many pastures into new cropland and were more interested in rising land prices than water conservation.

SPRINKLER IRRIGATION

In the sprinkler method of irrigation, water is sprayed into the air and allowed to fall on the ground surface somewhat resembling rainfall. The spray is developed by the flow of water under pressure through small orifices or nozzles. The pressure is usually obtained by pumping. With careful selection of nozzle sizes, operating pressure and sprinkler spacing the amount of irrigation water required to refill the crop root zone can be applied nearly uniform at the rate to suit the infiltration rate of soil.

Advantages of Sprinkler Irrigation

- Elimination of the channels for conveyance, therefore no conveyance loss.
- Suitable to all types of soil except heavy clay.
- Suitable for irrigating crops where the plant population per unit area is very high. It is most suitable for oil seeds and other cereal and vegetable crops.
- Water saving.
- Closer control of water application convenient for giving light and frequent irrigation and higher water application efficiency.
- Increase in yield.
- Mobility of system.
- May also be used for undulating area.
- Saves land as no bunds etc. are required.
- Influences greater conducive micro-climate.
- Areas located at a higher elevation than the source can be irrigated.
- Possibility of using soluble fertilizers and chemicals.
- Less problem of clogging of sprinkler nozzles due to sediment laden water.

Crop Response to Sprinkler

The trials conducted in different parts of the country revealed water saving due to sprinkler system varies from 16 to 70 % over the traditional method with yield increase from 3 to 57 % in different crops and agro climatic conditions.

Table: Response of Different Crops to Sprinkler Irrigation.

Crops	Water Saving, %	Yield increase, %
Bajra	56	19
Barley	56	16
Bhindi	28	23
Cabbage	40	3
Cauliflower	35	12
Chillies	33	24
Cotton	36	50
Cowpea	19	3
Fenugreek	29	35
Garlic	28	6
Gram	69	57
Groundnut	20	40
Jowar	55	34
Lucerne	16	27
Maize	41	36
Onion	33	23
Potato	46	4
Sunflower	33	20
Wheat	35	24

General Classification of Different Types of Sprinkler Systems

Sprinkler systems are classified into the following two major types on the basis of the arrangement for spraying irrigation water.

1) Rotating head or revolving sprinkler system:

Small size nozzles are placed on riser pipes fixed at uniform intervals along the length of the lateral pipe and the lateral pipes are usually laid on the ground surface. They may also be mounted on posts above the crop height and rotated through 90°, to irrigate a rectangular strip. In rotating type sprinklers, the most common device to rotate the sprinkler heads is with a small hammer activated by the thrust of water striking against a vane connected to it.

Example of a few rotating type sprinkler irrigation systems.

2) Perforated pipe system:

This method consists of drilled holes or nozzles along their length through which water is sprayed under pressure. This system is usually designed for relatively low pressure (1 kg/cm^2). The application rate ranges from 1.25 to 5 cm per hour for various pressure and spacing. Based on the portability, sprinkler systems are classified into the following types:

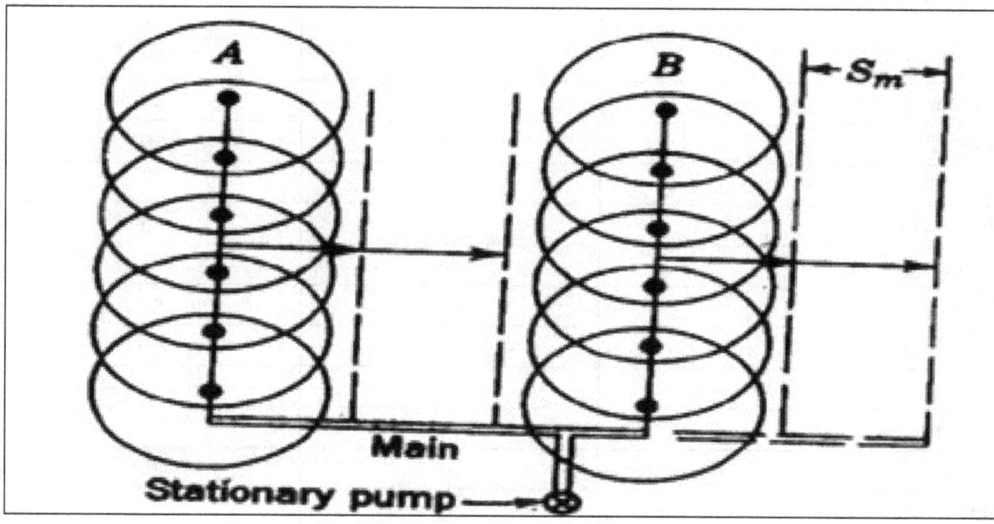

Fully portable sprinkler irrigation system.

- Portable system: A portable system has portable main lines, laterals and pumping plan. Semi portable system is similar to a portable system except that the location of water source and pumping plant is fixed.
- Semi permanent system: A semi permanent system has portable lateral lines, permanent main lines and sub mains and a stationery water source and pumping plant.
- Solid set system: A solid set system has enough laterals to eliminate their movement. The laterals are positions in the field early in the crop season and remain for the season.
- Permanent system: A fully permanent system consists of permanently laid mains, sub mains and laterals and a stationery water source and pumping plant.

Components of Sprinkler Irrigation System

A sprinkler system usually consists of the following components:

- A pump unit,
- Tubings- main/submains and laterals,
- Couplers,
- Sprinker head,
- Other accessories such as valves, bends, plugs and risers.

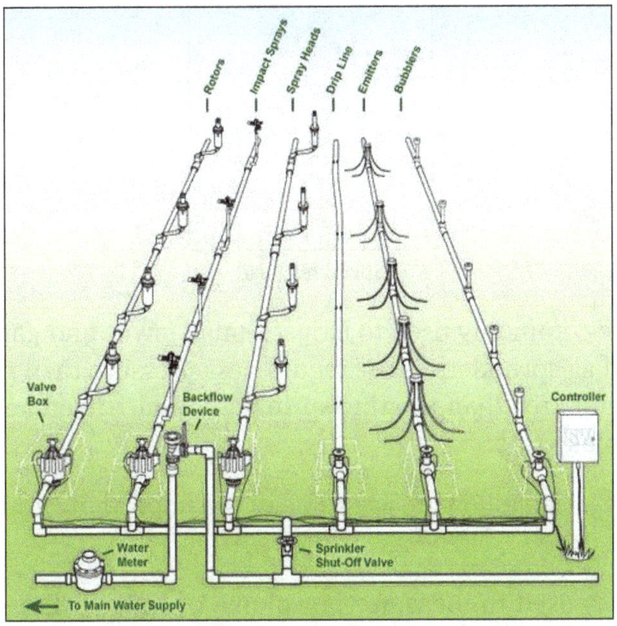

Component of a portable sprinkler irrigation system.

1. Pumping Unit: Sprinkler irrigation systems distribute water by spraying it over the fields. The water is pumped under pressure to the fields. The pressure forces the water through sprinklers or through perforations or nozzles in pipelines and then forms a spray. A high speed centrifugal or turbine pump can be used for operating sprinkler irrigation for individual fields. Centrifugal pump is used when the distance from the pump inlet to the water surface is less than eight meters. For pumping water from deep wells or more than eight meters, a turbine pump is suggested. The driving unit may be either an electric motor or an internal combustion engine.

2. Tubings (Mains/submains and laterals): The tubings consist of mainline, submanins and laterals. Main line conveys water from the source and distributes it to the submains. The submains convey water to the laterals which in turn supply water to the sprinklers. Aluminum or PVC pipes are generally used for portable systems, while steel pipes are usually used for center-pivot laterals. Asbestos, cement, PVC and wrapped steel are usually used for buried laterals and main lines.

3. Couplers: Couplers are used for connecting two pipes and uncoupling quickly and easily. Essentially a coupler should provide:

- A reuse and flexible connection.
- Not leak at the joint.
- Be simple and easy to couple and uncouple.
- Be light, non-corrosive, durable.

4. Sprinkler Head: Sprinkler head distribute water uniformly over the field without runoff or excessive loss due to deep percolation. Different types of sprinklers are available. They are either rotating or fixed type. The rotating type can be adapted for a wide range of application rates and spacing. They are effective with pressure of about 10 to 70 m head at the sprinkler. Pressures ranging from 16 to 40 m head are considered the most practical for most farmers.

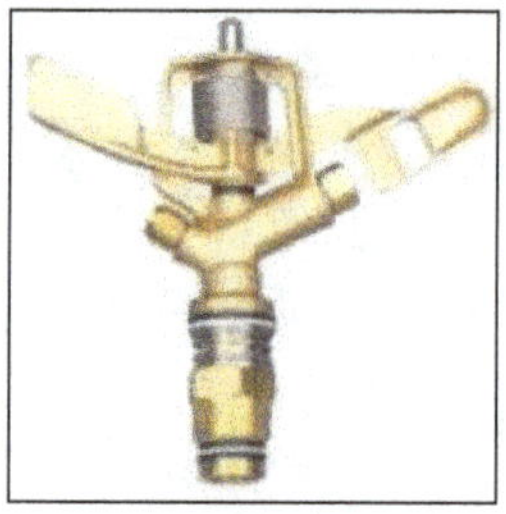

Sprinkler head

Fixed head sprinklers are commonly used to irrigate small lawns and gardens. Perforated lateral lines are sometimes used as sprinklers. They require less pressure than rotating sprinklers. They release more water per unit area than rotating sprinklers. Hence fixed head sprinklers are adaptable for soils with high intake rate.

5. Fittings and accessories: The following are some of the important fittings and accessories used in sprinkler system:

- Water meters: It is used to measure the volume of water delivered. This is necessary to operate the system to give the required quantity of water.
- Flange, couplings and nipple used for proper connection to the pump, suction and delivery.
- Pressure gauge: It is necessary to know whether the sprinkler system is working with desired pressure to ensure application uniformity.
- Bend, tees, reducers, elbows, hydrants, butterfly valve and plugs.
- Fertilizer applicator: Soluble chemical fertilizers can be injected into the sprinkler system and applied to the crop. The equipment for fertiliser application is relatively cheap and simple and can be fabricated locally. The fertilizer applicator consists of a sealed fertilizer tank with necessary tubings and connections. A venturi injector can be arranged in the main line, which creates the differential pressure suction and allows the fertilizer solution to flow in the main water line.

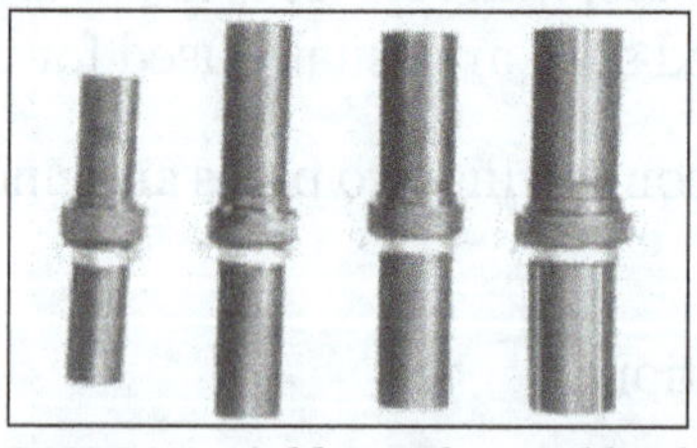

QRC HDPE sprinkler with metal latches.

QRC HDPE sprinkler with plastic latches.

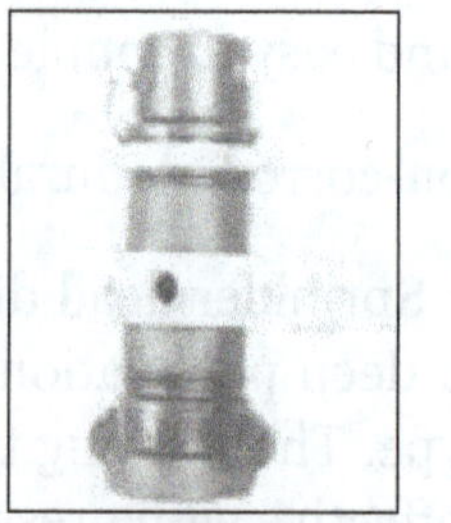

QRC HDPE sprinkler base.

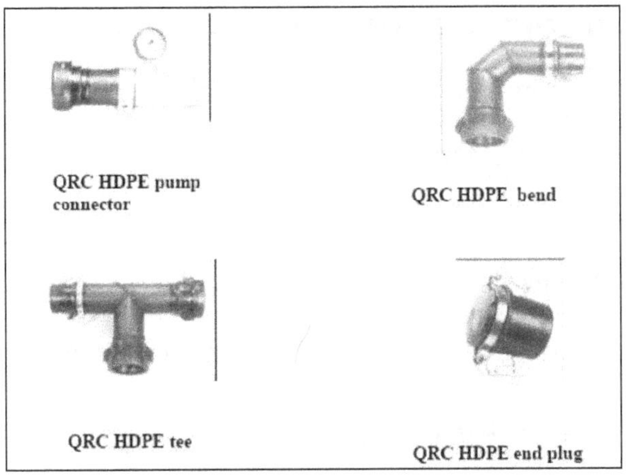

Different sprinkler pipes and fittings.

Rules for Sprinkler System Design

- Main should be laid up and down hill.
- Lateral should be laid across the slope or nearly on the contour.
- For multiple lateral operation, lateral pipe sizes should not be more than two diameter.
- Water supply source should be nearest to the center of the area.
- Layout should facilitate and minimize lateral movement during the season.
- Booster pump should be considered where small portion of field would require high pressure at the pump.
- Layout should be modified to apply different rates and amounts of water where soils are greatly different in the design area.

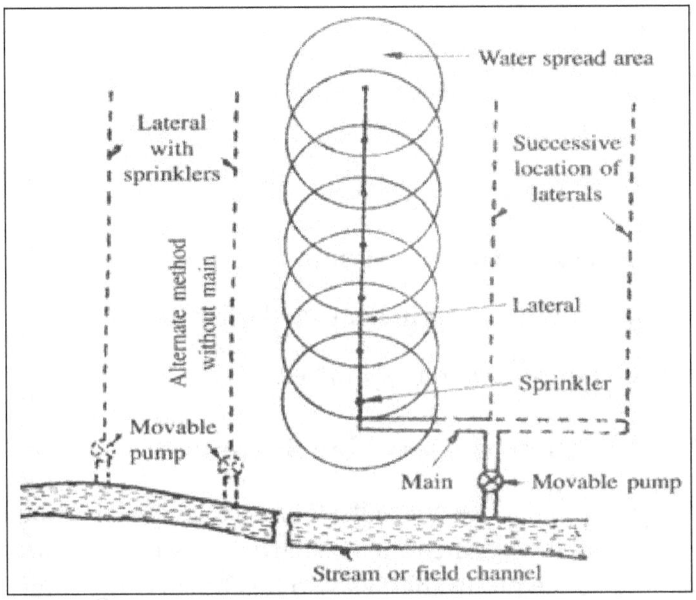

Layout of sprinkler irrigation system.

Selecting the most Appropriate Sprinkler Systems

While selecting a sprinkler system, the most important physical parameters to be considered are:

1. The crop or crops to be cultivated.
2. The shape and size (acres) of the field.
3. The topography of the field.
4. The amount of time and labor required to operate the system.

Selecting Sprinkler System Capacity

A sprinkler system must be designed to apply water uniformly without runoff or erosion. The application rate of the sprinkler system must be matched to the infiltration rate of the most restrictive soil in the field. If the application rate exceeds the soil intake rate, the water will runoff the field or relocate within the field resulting in over and under watered areas.

The sprinkler system capacity is the flow rate needed to adequately irrigate an area and is expressed in liters per minute per acre. The system capacity depends upon on the: Peak crop water requirements during the growing season; effective crop rooting depth; texture and infiltration rate of the soil; the available water holding capacity of the soil; pumping capacity of the well or wells (if wells are the water source).

Constraints in Application of Sprinkler Irrigation

1. Uneven water distribution due to high winds.
2. Evaporation loss when operating under high temperatures.
3. Highly impermeable soils are not suitable.
4. Initial cost is high.
5. Proper design.
6. Lack of Package of practices.
7. Lack of awareness.
8. Lack of social concern to save natural resources.
9. High water pressure required in sprinkler ($>2.5 kg/cm^2$).
10. Difficulty in irrigation during wind in sprinkler.

Operation and Maintenance of Sprinkler Systems

Proper design of a sprinkler system does not in itself ensure success. It should be ensured that the prime mover and the pump are in alignment, particularly in the case of tractor-driven pumps. For these the drive shaft as well as the pump shaft should lie at nearly the same height to prevent too great an angle on the universal shaft.

While laying the main and lateral pipes, always begin laying at the pump. This necessarily gives the correct connection of all quick coupling pipes. While joining couplings, it is ensured that both the couplings and the rubber seal rings are clean.

In starting the sprinkler system, the motor or engine is started with the valves closed. The pump must attain the pressure stated on type-plate or otherwise there is a fault in the suction line. After the pump reaches the regulation pressure, the delivery valve is opened slowly. Similarly, the delivery valve is closed after stopping the power unit.

The pipes and sprinkler-lines are shifted as required after stopping. Dismantling of the installation takes place in the reverse order to the assembly.

Maintenance

General principles regarding the maintenance of the pipes and fittings and sprinkler heads are given below:

1. Pipes and fittings: The pipes and fittings require virtually no maintenance but attention must be given to the following procedures:

- Occasionally clean any dirt or sand out of the groove in the coupler in which the rubber sealing ring fits. Any accumulation of dirt or sand will affect the performance of the rubber sealing ring.
- Keep all nuts and bolts tight.
- Do not lay pipes on new damp concrete or on piles of fertilizer. Do not lay fertilizer sacks on the pipe.

2. Sprinkler heads: The sprinkler heads should be given the following attention:

- When moving the sprinkler lines, make sure that the sprinklers are not damaged or pushed into the soil.
- Do not apply oil, grease or any lubricant to the sprinklers. They are water lubricated and using oil, grease or any other lubricant may stop them from working.
- Sprinklers usually have a sealed bearing and at the bottom of the bearing there are washers. Usually it is the washers that wear and not the more expensive metal parts.
- Check the washers for wear once a season or every six months which is especially important where water is sandy. Replace the washers if worn.
- After several season's operation the swing arm spring may need tightening. This is done by pulling out the spring end at the top and rebending it. This will increase the spring tension.

In general, check all equipment at the end of the season and make any repairs and adjustments and order the spare parts immediately so that the equipment is in perfect condition to start in the next season.

Storage

The following points are to be observed while storing the sprinkler equipment during the off season:

- Remove the sprinklers and store in a cool, dry place.
- Remove the rubber sealing rings from the couplers and fittings and store themin a cool, dark place.
- The pipes can be stored outdoors in which case they should be placed in racks with one end higher than the other. Do not store pipes along with fertilizer.
- Disconnect the suction and delivery pipe-work from the pump and pour in a small quantity of medium grade oil. Rotate the pump for a few minutes. Blank the suction and delivery branches. This will prevent the pump from rusting. Grease the shaft.
- Protect the electric motor from the ingress of dust, dampness and rodents.

Trouble Shooting

The following are the general guidelines to identify and remove the common troubles in the sprinkler systems:

- Pump does not prime or develop pressure:
    - Check that the suction lift is within the limits. If not, get the pump closer to the water.
    - Check the suction pipeline and all connections for air leaks. All connections and flanges should be air tight.
    - Check that the strainer on the foot valve is not blocked.
    - Check that the flap in the foot valve is free to open fully.
    - Check the pump gland for air leaks. If air leaks are suspected tighten the gland gently. If necessary repack the gland using a thick grease to seal the gland satisfactorily.
    - Check that the gate valve on the delivery pipe is fully closed during priming and opens fully when the pump is running.
- Sprinklers do not turn:
    - Check pressure.
    - Check that the nozzle is not blocked. Preferably unscrew the nozzle or use a small soft piece of wood to clear the blockage. Do not use a piece of wire or metal as this may damage the nozzle.
    - Check the condition of washers at the bottom of the bearing and replace them if worn or damaged.

- o Check that the swing arm moves freely and that the spoon which. moves into the water stream is not bent by comparing it with a sprinkler which is operating correctly.
- o Adjust the swing arm spring tension: Usually it should not be necessary to pull up the spring by more than about 6 mm.

- Leakage from coupler or fittings: The sealing rings in the couplers and fittings are usually designed to drain the water from the pipes when the pressure is turned off. This ensures that the pipes are automatically emptied and ready to be moved. With full pressure in the system the couplers and fittings will be effectively leak-free. If, however, there is a leakage, check the following:
    - o There is no accumulation of dirt or sand in the groove in the coupler in which the sealing ring fits. Clean out any dirt or sand and refit the sealing ring.
    - o The end of the pipe going inside the coupler is smooth, clean and not distorted.
    - o In the case of fittings such as bends, tees and reducers ensure that the fitting has been properly connected into the coupler.

Unit Cost of Sprinkler Systems

Different components required for a sprinkler irrigation system to irrigate 1 ha to 4 ha area and their cost has been estimated and given in table.

Table: Unit Cost of Sprinkler Systems.

Sprinkler system Components	Ha (50 mm dia)		2.0 ha (63 mm dia)		3.0 ha (75 mm dia)		4.0 ha (75 mm dia)	
	Quantity (No.)	Amount (Rs.)	Quantity (No.)	Amount (Rs.)	Quantity (No.)	Amount (Rs.)	Quantity (No.)	Amount (Rs.)
HDPE Pipes with quick action coupler (2.5 kg/cm2) of 6m long	25	7770.00	30	10448.7	37	13378.09	45	16270.65
Sprinkler coupler with foot baton assembly	5	921.60	7	1411.20	11	2382.30	14	3032.00
Sprinkler nozzles (1.7 to 2.8 kg/cm2)	5	1188.00	7	1662.20	11	2613.60	14	3326.40
Riser pipe 20mm diameter x 75cm long	5	264.00	7	369.60	11	580.80	14	739.20
Connecting nipple	1	115.20	1	129.00	1	167.00	1	167.00
Bend with coupler 900	1	108.00	1	126.00	1	168.00	1	168.00
Tee with coupler	1	138.00	1	152.40	1	180.00	1	180.00
End plug	2	96.00	2	108.00	2	132.00	2	132.00
Basic system cost per hectare (Rs.)		10600.80		14407.1		19601.79		24015.25

MICRO-IRRIGATION

Micro-irrigation, also called localised irrigation, low volume irrigation, low-flow irrigation, or trickle irrigation is an irrigation method with lower pressure and flow than a traditional sprinkler system. Low volume irrigation is used in agriculture for row crops, orchards, and vineyards. It is also used in horticulture in wholesale nurseries, in landscaping for civic, commercial, and private landscapes and gardens, and in the science and practice of restoration ecology and environmental remediation.

There are several types of micro-irrigation systems. Many of the components are the same for all of these types of systems. Most systems typically include filters, pipes, valves, and tubing. The main difference is in the type of emission device that is used to deliver the water to the plants. Drip irrigation utilizes drip emitters that deliver water at very low rates. The typical range is 0.2 to 4.0 gallons per hour. In some systems, the emitters are installed manually on the outside of the tubing and placed where needed. Other systems might use integral dripperline or drip tape with the emitters already installed at a predetermined spacing. Micro-sprinklers, which can include fixed stream sprays and rotating spinners typically deliver water at a higher rate, such as 10 to 25 gallons per hour and will cover a larger area than drip emitters. These are more typically used in tree orchards where the plants are larger. The goal is to distribute water slowly in small volumes and target it to plants' root zones with less runoff or overspray than landscape and garden conventional spray and rotary sprinklers. The low volume allows the water to penetrate and be absorbed into slow-percolation soils, such as clay, minimizing water runoff.

System Components

There are a wide variety of system components included in a micro-irrigation systems. Most systems include a filter. These may include pre-filters, sand separators, media filters, screen filters, and disc filters. The level of filtration required depends on the size of the emission device and the quality of the water source. A pressure regulator or regulating valve may be required to reduce the system pressure to the desired level. Automatic or manually operated valves will be required to switch from one irrigated section to another. An irrigation controller will be used with automatic systems and may also be needed for backflushing the filter or sand separator. Since water conservation is a frequent reason for choosing micro-irrigation systems, soil moisture sensors, rain shutoff sensors, and sometimes even weather stations may be installed.

Emission Devices

Microtubing

Microtubing is one of the oldest types of drip irrigation devices and was used in greenhouses in the 1970s. It consists of a very small diameter tubing. Flow is regulated purely by the length and diameter of the tubing. Weights or stakes are sometimes attached to the end of the tubing to keep it in place.

Fixed Flow Drip Emitters

Low-flow irrigation systems in gardens using drip apply water through two methods:

- Pre installed small holes in small diameter tubes placed on or below the surface.
- Self cleaning emitters, in different precipitation rates, pre installed or contractor installed for different rate emitters on same supply line (i.e. trees-higher, perennials-lower). The flexible supply pipe can be buried either underground or pinned on the surface and buried under.

Low volume irrigation systems often use the two delivery components of drip systems to apply water through small holes in small diameter tubes placed on or below the surface of the field. This is done instead of agricultural surface irrigation and furrow irrigation for vegetables, fruits and berries, and other high-value crops.

Adjustable Drip Emitters

Trickle emitters, also called 'spider sprays,' come in fixed or adjustable radius shapes and diameters, and are installed directly on the flexible supply pipe or on tubing connected to it, and mounted on small stakes. Trickle emitter-'Spider sprays' work well for plants with more fibrous root systems, tree and large shrub basins, and in pots and container gardens - allowing automated watering of plants on decks and patios. Mist emitters can also be used in pot, both on the ground and hanging, with humidity-fog watering for epiphytes and ferns replicating habitats.

In the Horticulture industry, wholesale growers and plant nurseries often use the trickle emitters for 5-US-gallon (19 L) and larger container stock, to automate watering. Attached to longer supply tubing on short stakes, they are easily movable to new containers when stock is moved or sold. Mist emitters are used for propagation, epiphytes, and other plants needing higher humidity.

Micro-sprinklers

Low volume micro-sprinklers may be attached to hard plastic risers or attached to standard sprinkler heads, but are more typically mounted on stakes and attached to small diameter micro-tubing connected to polyethylene tubing with a barbed connector. Some micro-sprinklers have a fixed spray or stream pattern, while others rotate. These are installed above ground and are often used for fruit and nut orchards and vineyards. These systems are expensive, even for large-scale agricultural use, and are predominantly used for high-value crops.

Macro-drip Irrigation

High-volume, low-pressure irrigation systems for container gardening are known as Macro-Drip. A pressure regulator lowers the water pressure to under 30 pounds per square inch while a relatively large diameter hose or pipe delivers the water directly to a sprinkler head. This allows a larger volume of water to reach the flowerpot in a short amount of time, which will then be absorbed into the roots of the plant.

Ecological Restoration and Phytoremediation Projects

Low-flow irrigation systems are used on some native plant habitat restoration and environmental remediation projects. The lower operating pressure can be the only choice for remote locations with wells or small storage tank water sources. It is used in temporary installations during initial establishment periods, and being on the soil surface easily removable with minimal damage to the recovering plant community. An example is its use in riparian zone restoration, and environmental remediation projects using Phytoremediation and Bioremediation techniques.

Water Conservation and Regulations

As municipal and agricultural water supplies become more limited; through increased population demands, droughts, and climate change; city, water district, and state-province level regulations and codes are beginning to encourage, offer rebates with use, or mandate significantly reduced water allowances, at higher costs, that are bringing many water conservation products and techniques both to the forefront and more competitively matched to traditional irrigation system costs.

Use of micro-irrigation systems on green building candidate projects can help them to accumulate points for LEED - (Leadership in Energy and Environmental Design) certification rating and awards.

DRIP IRRIGATION

Drip irrigation is a type of micro-irrigation system that has the potential to save water and nutrients by allowing water to drip slowly to the roots of plants, either from above the soil surface or buried below the surface. The goal is to place water directly into the root zone and minimize evaporation. Drip irrigation systems distribute water through a network of valves, pipes, tubing, and emitters. Depending on how well designed, installed, maintained, and operated it is, a drip irrigation system can be more efficient than other types of irrigation systems, such as surface irrigation or sprinkler irrigation.

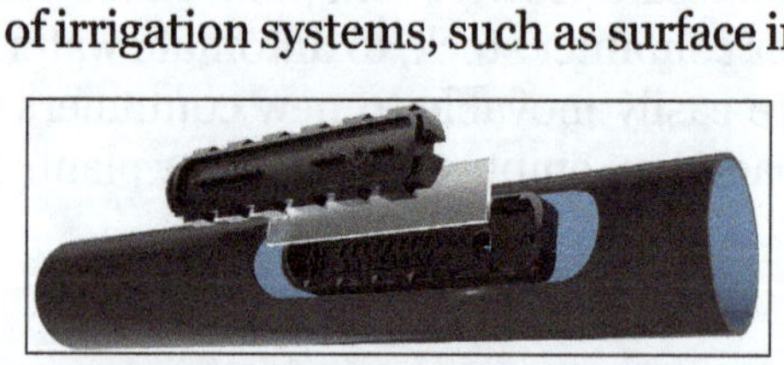

Open pressure compensated dripper.

Components and Operation

Components used in drip irrigation (listed in order from water source) include:

- Pump or pressurized water source.
- Water filter or filtration systems: sand separator, Fertigation systems (Venturi injector) and chemigation equipment (optional).
- Backwash controller (Backflow prevention device).
- Pressure Control Valve (pressure regulator).

- Distribution lines (main larger diameter pipe, maybe secondary smaller, pipe fittings).
- Hand-operated, electronic, or hydraulic control valves and safety valves.
- Smaller diameter polyethylene tube (often called "laterals").
- Poly fittings and accessories (to make connections).
- Emitting devices at plants (emitter or dripper, micro spray head, inline dripper or inline drip tube).

In drip irrigation systems, pump and valves may be manually or automatically operated by a controller.

Most large drip irrigation systems employ some type of filter to prevent clogging of the small emitter flow path by small waterborne particles. New technologies are now being offered that minimize clogging. Some residential systems are installed without additional filters since potable water is already filtered at the water treatment plant. Virtually all drip irrigation equipment manufacturers recommend that filters be employed and generally will not honor warranties unless this is done. Last line filters just before the final delivery pipe are strongly recommended in addition to any other filtration system due to fine particle settlement and accidental insertion of particles in the intermediate lines.

Drip and subsurface drip irrigation is used almost exclusively when using recycled municipal wastewater. Regulations typically do not permit spraying water through the air that has not been fully treated to potable water standards.

Because of the way the water is applied in a drip system, traditional surface applications of timed-release fertilizer are sometimes ineffective, so drip systems often mix liquid fertilizer with the irrigation water. This is called fertigation; fertigation and chemigation (application of pesticides and other chemicals to periodically clean out the system, such as chlorine or sulfuric acid) use chemical injectors such as diaphragm pumps, piston pumps, or aspirators. The chemicals may be added constantly whenever the system is irrigating or at intervals. Fertilizer savings of up to 95% are being reported from recent university field tests using drip fertigation and slow water delivery as compared to timed-release and irrigation by micro spray heads.

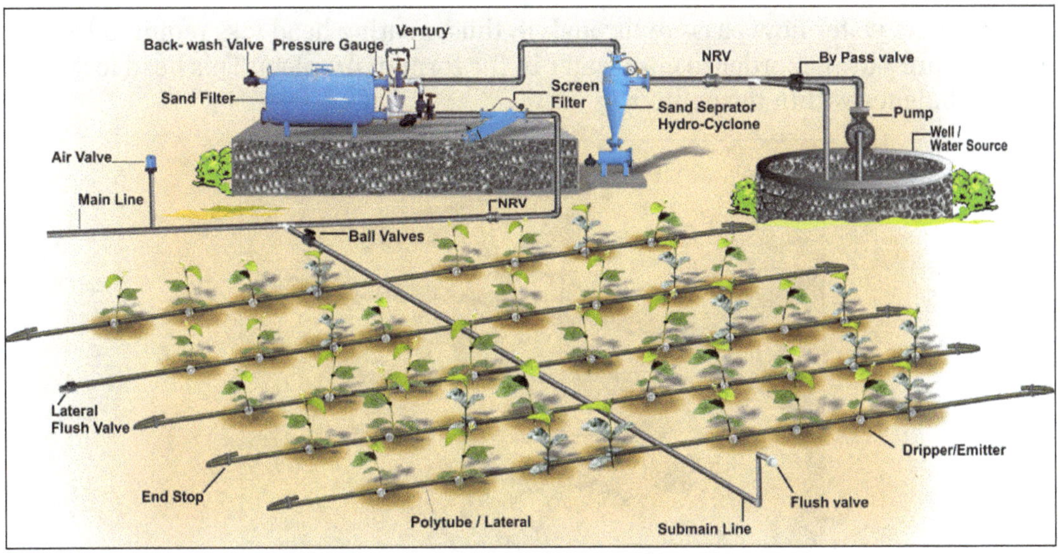

Drip irrigation system layout and its parts.

Nursery flowers watered with drip irrigation in Israel.

Properly designed, installed, and managed, drip irrigation may help achieve water conservation by reducing evaporation and deep drainage when compared to other types of irrigation such as flood or overhead sprinklers since water can be more precisely applied to the plant roots. In addition, drip can eliminate many diseases that are spread through water contact with the foliage. Finally, in regions where water supplies are severely limited, there may be no actual water savings, but rather simply an increase in production while using the same amount of water as before. In very arid regions or on sandy soils, the preferred method is to apply the irrigation water as slowly as possible.

Pulsed irrigation is sometimes used to decrease the amount of water delivered to the plant at any one time, thus reducing runoff or deep percolation. Pulsed systems are typically expensive and require extensive maintenance. Therefore, the latest efforts by emitter manufacturers are focused on developing new technologies that deliver irrigation water at ultra-low flow rates, i.e. less than 1.0 liter per hour. Slow and even delivery further improves water use efficiency without incurring the expense and complexity of pulsed delivery equipment.

An emitting pipe is a type of drip irrigation tubing with emitters pre-installed at the factory with specific distance and flow per hour as per crop distance.

An emitter restricts water flow passage through it, thus creating head loss required (to the extent of atmospheric pressure) in order to emit water in the form of droplets. This head loss is achieved by friction/turbulence within the emitter.

Horticulture drip emitter in a pot.

Advantages and Disadvantages

Pot irrigation by On-line drippers.

Drip irrigation and spare drip irrigation tubes in banana farm.

Pressure compensated integral dripper on soilless without growing channels.

The Advantages of Drip Irrigation are:

- Fertilizer and nutrient loss is minimized due to a localized application and reduced leaching.
- Water application efficiency is high if managed correctly.
- Field leveling is not necessary.
- Fields with irregular shapes are easily accommodated.
- Recycled non-potable water can be safely used.
- Moisture within the root zone can be maintained at field capacity.
- Soil type plays a less important role in the frequency of irrigation.
- Soil erosion is lessened.
- Weed growth is lessened.
- Water distribution is highly uniform, controlled by the output of each nozzle.
- Labour cost is less than other irrigation methods.
- Variation in supply can be regulated by regulating the valves and drippers.
- Fertigation can easily be included with minimal waste of fertilizers.
- Foliage remains dry, reducing the risk of disease.
- Usually operated at lower pressure than other types of pressurized irrigation, reducing energy costs.

The Disadvantages of Drip Irrigation are:

- Initial cost can be more than overhead systems.
- The sun can affect the tubes used for drip irrigation, shortening their lifespan.
- The risks of degrading plastic affecting the soil content and food crops. With many types of plastic, when the sun degrades the plastic, causing it to become brittle, the estrogenic chemicals (that is, chemicals replicating female hormones) which would cause the plastic to retain flexibility have been released into the surrounding environment.
- If the water is not properly filtered and the equipment not properly maintained, it can result in clogging or bioclogging.
- For subsurface drip the irrigator cannot see the water that is applied. This may lead to the farmer either applying too much water (low efficiency) or an insufficient amount of water, this is particularly common for those with less experience with drip irrigation.
- Drip irrigation might be unsatisfactory if herbicides or top dressed fertilizers need sprinkler irrigation for activation.
- Drip tape causes extra cleanup costs after harvest. Users need to plan for drip tape winding, disposal, recycling or reuse.

- Waste of water, time and harvest, if not installed properly. These systems require careful study of all the relevant factors like land topography, soil, water, crop and agro-climatic conditions, and suitability of drip irrigation system and its components.
- In lighter soils subsurface drip may be unable to wet the soil surface for germination. Requires careful consideration of the installation depth.
- Most drip systems are designed for high efficiency, meaning little or no leaching fraction. Without sufficient leaching, salts applied with the irrigation water may build up in the root zone, usually at the edge of the wetting pattern. On the other hand, drip irrigation avoids the high capillary potential of traditional surface-applied irrigation, which can draw salt deposits up from deposits below.
- The PVC pipes often suffer from rodent damage, requiring replacement of the entire tube and increasing expenses.
- Drip irrigation systems cannot be used for damage control by night frosts (like in the case of sprinkler irrigation systems).

Drip Tape

Drip tape is a type of thin-walled dripperline used in drip irrigation. The first drip tape was known as "Dew Hose".

Drip tape duct tape is made of polyethylene and is sold flat on reels. The wall thickness typically ranges from 4 to 25 mils (0.1–0.6 mm). Thicker walled tapes are commonly used for permanent subsurface drip irrigation and thinner walled tapes for temporary throw-away type systems in high-value crops.

Water exits from tape through emitters or drippers. The typical emitter spacing ranges from 6 to 24 inches (150–600 mm). In some products, the emitters are manufactured simultaneously with the tape and are actually formed as part of the product itself. In others, the emitters are manufactured separately and installed at the time of production.

Some product is not a tape, but a thin-walled dripperline, but in popular parlance, both types of products are called tapes. Typical tape diameters are 5/8", 7/8", and 1-3/8", with the larger diameters more commonly used on permanent installations with longer runs.

Drip tape is a recyclable material and can be recycled into viable plastic resins for reuse in the plastics manufacturing industry.

Uses

Irrigation dripper.

Drip irrigation is used in farms, commercial greenhouses, and residential gardens. Drip irrigation is adopted extensively in areas of acute water scarcity and especially for crops and trees such as coconuts, containerized landscape trees, grapes, bananas, ber, eggplant, citrus, strawberries, sugarcane, cotton, maize, and tomatoes.

Drip irrigation for garden available in drip kits are increasingly popular for the homeowner and consist of a timer, hose and emitter. Hoses that are 4 mm in diameter are used to irrigate flower pots.

PULSE DRIP IRRIGATION

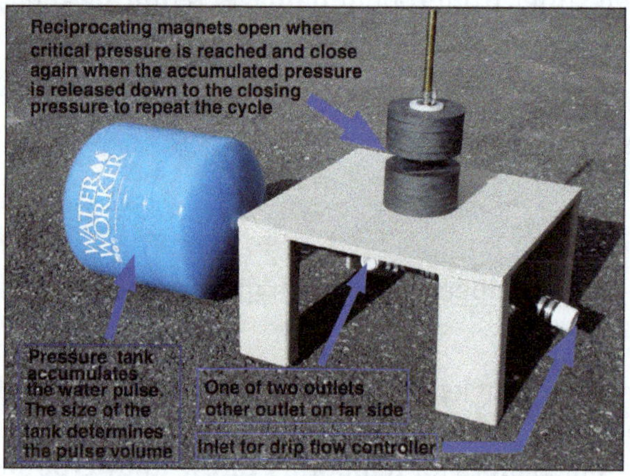

Illustration of a pulse drip irrigation system.

Magnets hold the valve closed while water comes in through the inlet and drip flow controller to charge the reservoir. When the building pressure finally overcomes the attractive force of the magnets the top set of magnets are forced upward to open the valve. As the water is released, the pressure decreases to a point where the magnets drop back down and the charging phase begins anew.

In the above figure is shown, one of two pulsing impact sprinkler at The Crescent Grange in Broomfield, Colorado providing 5,000 square feet (460 m²) of coverage of their permaculture garden. The sprinklers were run continuously, delivering 5-second pulses per minute for the first twelve days to germinate the seeds. Thereafter, it was gradually cut back to 8- to 10-hour watering periods three times a week.

Pulse drip irrigation is an experimental irrigation technique primarily used with drip irrigation. Maintaining a high level of soil moisture for germination of seed is one reason this technique may be used.

Most conventional drip irrigation systems can be made to pulse by using a timer to reduce the watering duration and increase the watering frequency. Some newer systems have been developed that utilize a pressurized reservoir. When the pressure in the reservoir reaches some predetermined pressure level the valve on the reservoir opens and a portion of the fluid contained within the reservoir is forcefully discharged. While the fluid is discharging, the pressure within the reservoir decreases. When the decrease in water pressure reaches a predetermined level the valve closes to resume the charging phase. The charge-discharge cycling will continue as long as the flow rate coming in through the inlet is less than the expel rate passing out through the outlets while the valve is open. A device called a drip flow controller is placed at the inlet for this purpose to regulate the flow into the inlet.

If properly designed and operated, a low-flow pulse system may be left operating continuously for a period of time without overwatering. Constant and frequent irrigation applications have been cited as one way to reduce water demand. Some literature also cite the benefits of small frequent watering applications to reduce water stress on plants.

Landscaping in sand with *in situ* seed germination. A one-US-gallon-per-hour (3.8 L) drip flow controller feeding an 82-foot-long (25 m) drip line with check valves comprising 82 drip points along its length so each drip point is putting out about 1/82 US gallon (46 ml) per hour.

Crimson clover sprouts grown on 1/8-inch (0.32 cm) urethane foam mats and flagstone. Four micro-sprinklers cycle pulsing continuously over a 7-day period, each putting out about 1/2 US gallon (1.9 l) per hour. The four micro-sprinklers were each fitted with an LPD to keep the lines fully charged between pulses.

Low-flow application rates can be used with different soils and growth media. The water can be applied slowly enough to match the water infiltration rate and prevent water loss from deep percolation or runoff. Mineral nutrients added to media with a high void content, such as coarse grained sand, will provide more oxygen to roots than ordinary soil and share some of the advantages with aeroponics. Sand also has a low water retention potential that makes it easier for plants to extract water by expending less energy due to the sand's relatively large particle size, which consequently does not bind very well to water. This increases the plant's water-use efficiency. Sand is also less hospitable to pathogens that can attack roots.

Advantages

Pulse drip irrigation has the following advantages:

- Less run off on heavy soils.
- Less leaching or water loss in sandy soils.
- Water can be applied more efficiently on shallow soils and in hilly areas.
- Intermittent operation of sprinklers and foggers can provide evaporative cooling for temperature control.
- Size of growing containers in greenhouses may be reduced due to very low discharge rates.
- Low flow inputs may reduce supply system costs because smaller pipes and lower capacity pumps would be needed.

Disadvantages

- The additional expense of buying and maintaining a pulse drip system to a pre-existing irrigation system.
- The system requires a minimum of 170 kPa (25 psi) water pressure to run efficiently.
- Maintaining the integrity of a pressurized water supply is of critical importance. Leaks in the piping can run up substantial costs due to the long duration of the watering periods.
- Close attention to the watering cycles is needed to avoid salt build up in the soil. Light watering can leave salt residues behind as the water evaporates and leave little added moisture to the soil.
- The additional expense of leak prevention devices (LPDs) or check valves are recommended to be fitted at each watering distribution point along the system to ensure even distribution of water. If the lines are allowed to drain between pulses, most of the water will preferentially flow to those at the beginning of the line.

DEFICIT IRRIGATION

Deficit irrigation (DI) is a watering strategy that can be applied by different types of irrigation application methods. The correct application of DI requires thorough understanding of the yield response to water (crop sensitivity to drought stress) and of the economic impact of reductions in harvest. In regions where water resources are restrictive it can be more profitable for a farmer to maximize crop water productivity instead of maximizing the harvest per unit land. The saved water can be used for other purposes or to irrigate extra units of land. DI is sometimes referred to as incomplete supplemental irrigation or regulated DI.

Deficit irrigation is an optimization strategy in which irrigation is applied during drought-sensitive growth stages of a crop. Outside these periods, irrigation is limited or even unnecessary if rainfall provides a minimum supply of water. Water restriction is limited to drought-tolerant phenological stages, often the vegetative stages and the late ripening period. Total irrigation application is therefore not proportional to irrigation requirements throughout the crop cycle. While this inevitably results in plant drought stress and consequently in production loss, DI maximizes irrigation water productivity, which is the main limiting factor. In other words, DI aims at stabilizing yields and at obtaining maximum crop water productivity rather than maximum yields.

Crop Water Productivity

Crop water productivity (WP) or water use efficiency (WUE) expressed in kg/m^3 is an efficiency term, expressing the amount of marketable product (e.g. kilograms of grain) in relation to the amount of input needed to produce that output (cubic meters of water). The water used for crop production is referred to as crop evapotranspiration. This is a combination of water lost by evaporation from the soil surface and transpiration by the plant, occurring simultaneously. Except by modeling, distinguishing between the two processes is difficult. Representative values of WUE for cereals at field level, expressed with evapotranspiration in the denominator, can vary between 0.10 and 4 kg/m^3.

Experiences with Deficit Irrigation

For certain crops, experiments confirm that DI can increase water use efficiency without severe yield reductions. For example for winter wheat in Turkey, planned DI increased yields by 65% as compared to winter wheat under rainfed cultivation, and had double the water use efficiency as compared to rainfed and fully irrigated winter wheat. Similar positive results have been described for cotton. Experiments in Turkey and India indicated that the irrigation water use for cotton could be reduced to up to 60 percent of the total crop water requirement with limited yield losses. In this way, high water productivity and a better nutrient-water balance was obtained.

Certain Underutilized and horticultural crops also respond favorably to DI, such as tested at experimental and farmer level for the crop quinoa. Yields could be stabilized at around 1.6 tons per hectare by supplementing irrigation water if rainwater was lacking during the plant establishment and reproductive stages. Applying irrigation water throughout the whole season (full irrigation) reduced the water productivity. Also in viticulture and fruit tree cultivation, DI is practiced.

Scientists affiliated with the Agricultural Research Service (ARS) of the USDA found that conserving water by forcing drought (or deficit irrigation) on peanut plants early in the growing season has shown to cause early maturation of the plant yet still maintain sufficient yield of the crop. Inducing drought through deficit irrigation earlier in the season caused the peanut plants to physiologically "learn" how to adapt to a stressful drought environment, making the plants better able to cope with drought that commonly occurs later in the growing season. Deficit irrigation is beneficial for the farmers because it reduces the cost of water and prevents a loss of crop yield (for certain crops) later on in the growing season due to drought. In addition to these findings, ARS scientists suggest that deficit irrigation accompanied with conservation tillage would greatly reduce the peanut crop water requirement.

For other crops, the application of deficit irrigation will result in a lower water use efficiency and yield. This is the case when crops are sensitive to drought stress throughout the complete season, such as maize.

Reasons for Increased Water Productivity under Deficit Irrigation

If crops have certain phenological phases in which they are tolerant to water stress, DI can increase the ratio of yield over crop water consumption (evapotranspiration) by either reducing the water loss by unproductive evaporation, and by increasing the proportion of marketable yield to the totally produced biomass (harvest index), and by increasing the proportion of total biomass production to transpiration due to hardening of the crop - although this effect is very limited due to the conservative relation between biomass production and crop transpiration, - and due to adequate fertilizer application and by avoiding bad agronomic conditions during crop growth, such as water logging in the root zone, pests and diseases, etc.

Advantages

The correct application of DI for a certain crop:

- Maximizes the productivity of water, generally with adequate harvest quality.
- Allows economic planning and stable income due to a stabilization of the harvest in comparison with rainfed cultivation.
- Decreases the risk of certain diseases linked to high humidity (e.g. fungi) in comparison with full irrigation.
- Reduces nutrient loss by leaching of the root zone, which results in better groundwater quality and lower fertilizer needs as for cultivation under full irrigation.
- Improves control over the sowing date and length of the growing period independent from the onset of the rainy season and therefore improves agricultural planning.

Constraints

A number of constraints apply to deficit irrigation:

- Exact knowledge of the crop response to water stress is imperative.
- There should be sufficient flexibility in access to water during periods of high demand (drought sensitive stages of a crop).
- A minimum quantity of water should be guaranteed for the crop, below which DI has no significant beneficial effect.

- An individual farmer should consider the benefit for the total water users community (extra land can be irrigated with the saved water), when he faces a below-maximum yield.
- Because irrigation is applied more efficiently, the risk for soil salinization is higher under DI as compared to full irrigation.

Field experimentation is necessary for correct application of DI for a particular crop in a particular region. In addition, simulation of the soil water balance and related crop growth (crop water productivity modeling) can be a valuable decision support tool. By conjunctively simulating the effects of different influencing factors (climate, soil, management, crop characteristics) on crop production, models allow to (1) better understand the mechanism behind improved water use efficiency, to (2) schedule the necessary irrigation applications during the drought sensitive crop growth stages, considering the possible variability in climate, to (3) test DI strategies of specific crops in new regions, and to (4) investigate the effects of future climate scenarios or scenarios of altered management practices on crop production.

SUBSURFACE TEXTILE IRRIGATION

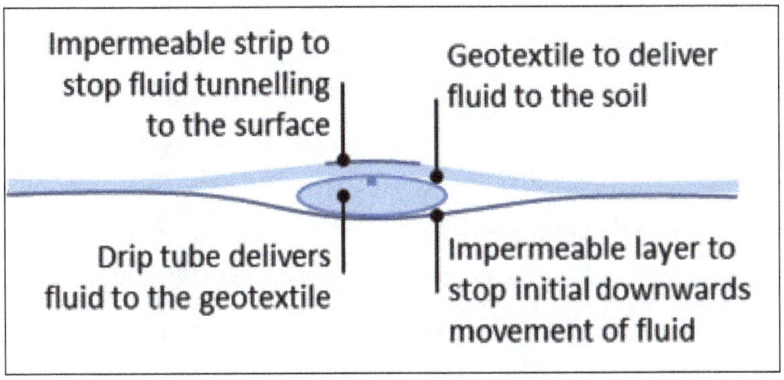

Diagram showing the structure of an example SSTI installation.

Subsurface Textile Irrigation (SSTI) is a technology designed specifically for subsurface irrigation in all soil textures from desert sands to heavy clays. Use of SSTI will significantly reduce the usage of water, fertilizer and herbicide. It will lower on-going operational costs and, if maintained properly, will last for decades. By delivering water and nutrients directly to the root zone, plants are healthier and have a far greater yield.

It is the only irrigation system that can safely use recycled water or treated water without expensive "polishing" treatment because water never reaches the surface.

A typical subsurface textile irrigation system has an impermeable base layer (usually polyethylene or polypropylene), a drip line running along that base, a layer of geotextile on top of the drip line and, finally, a narrow impermeable layer on top of the geotextile. Unlike standard drip irrigation, the spacing of emitters in the drip pipe is not critical as the geotextile moves the water along the fabric up to 2m from the dripper.

SSTI is installed 15–20 cm below the surface for residential/commercial applications and 30–50 cm for agricultural applications.

How SSTI Works

The systems rely on specific geotextiles to absorb the water from the drippers and to rapidly transport that water via mass flow and capillary action along the geotextile effectively turning those single drippers into billions of emitters. In essence, this enables intimate control of the speed of water delivery so that the capillary action of any soil can be matched (something that is virtually impossible for any other irrigation method including bare drip pipe below the surface). If the capillary action of the soil can be matched with the water delivery, only the minimal amount of water is needed to service the needs of the plants above (note that capillary forces draw water up from the water source to the roots of the plants).

To increase effectiveness, SSTI products should have an impermeable base layer to slow gravitational loss of water and to create an elliptical wetting pattern under the soil surface. It should also have a small impermeable top layer to ensure that water from the dripper does not "tunnel" through the geotextile and up to the surface (again a common problem with bare subsurface drip pipe). The effect of these two layers is dramatic as it maximizes the spread of water through the geotextile (up to 10,000 times faster than a clay loam soil as tested).

When comparing SSTI with surface drip, using the same amount of water, SSTI can cover 2.5 times the volume of soil and takes six times longer to dry down until the next irrigation is required.

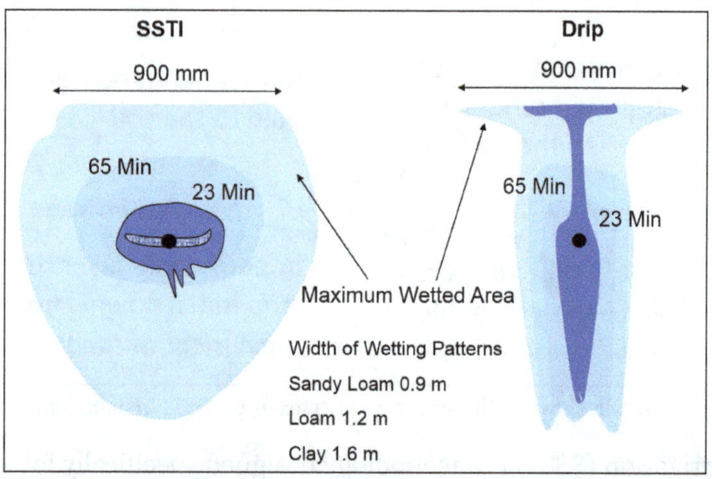

A cross-sectional view of the wetting pattern provided by SSTI, as compared to drip irrigation.

Recycled Water and Treated Effluent

Recycled water can be used in SSTI systems as it will spread the nutrient load over 2-3 times the soil volume (compared to other irrigation methods). This means that additional nutrient requirements are minimized and the soil will have a long life without overloading other nutrients (especially phosphorus and potassium).

A major benefit of SSTI is that treated effluent can be used but is prevented from reaching the surface. Recreational or agricultural activities can continue on the field during irrigation without the contaminants coming in contact with the public.

Nutrients can be injected through all SSTI systems (fertigation). Macro and micro nutrients can be delivered to specific crops including grass, pasture, trees and vines. The nutrient is placed directly in the root zone so there is almost no wastage and no potential for run off into waterways.

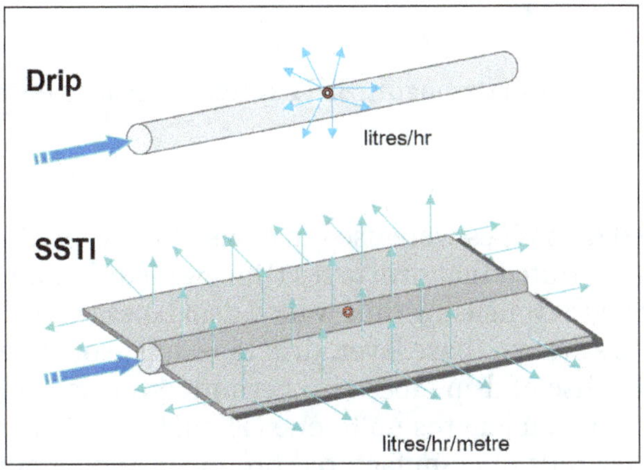

The flow of water through an SSTI installation as compared to a drip irrigation system.

Use of Geotextiles

Geotextiles were used for capillary irrigation not long after World War II. However, it was not until 1995 that extensive research was conducted at the CSIRO Division of Land and Water in Griffith, Australia, by Grain Security Foundation Ltd, that SSTI was established as a serious commercial alternative to drip.

Polyester geotextile of specific thickness and manufacture is required to ensure the system has the appropriate flow characteristics and so that it does not become hydrophobic (repels water).

Solving Typical Subsurface Problems

Studies were done on many forms of geotextile using various dripper rates and configurations to evaluate water flow for the major soil texture types and to determine if SSTI had the same problems experienced by subsurface drip systems (SDI). Specifically, considerable focus was placed on the problem of root intrusion, tunnelling, dripper blockage and insect damage (serious problems associated with SDI).

Root Intrusion

Research data showed that roots could invade the geotextile but did not thrive nor thicken in the polyester geotextile and, therefore, caused no damage. The roots were deterred from entering the dripper area because that area is dry relative to the rest of the system. Roots simply went elsewhere.

Tunneling

Tunneling (the process whereby water works its way up to the surface) was nearly eliminated by the narrow reflective impermeable tape above the drip line.

Blocked Drip Emitters

Drippers are often blocked in bare drip systems due to soil lodging in the emitters and soil crusting around the emitters when the system is not operating. SSTI eliminates this problem because there is a physical barrier presented by the geotextile and due to the fact that the soil remains moist for much longer than drip systems (i.e. the soil does not form a crust anyway).

Insects

Similarly, insects are deterred from damaging SSTI systems because of the geotextile barrier.

Water Delivery

Initially solid, thick-walled drip pipes were used in trials at the CSIRO but subsequent trials between 1995 and 1998 demonstrated that drip tapes (thin-walled, flexible pipe) could be used very effectively. The loss of strength from using thin-walled drip tapes was countered by the additional tensile strength of the entire system (base layer, drip tape and geotextile). While significantly reducing the cost of SSTI, the use of drip tapes also permitted the use of large diameter drip lines (of up to 35mm) allowing for run lengths up to one (1) kilometre. The thin, flexible wall of drip tape also meant that master rolls of manufactured product could be much larger (over 600m in length).

Optimal Width and Wetting Patterns

The width of SSTI products varies based on the application. However, at the University of Queensland, Australia trials were conducted to discover the optimal width of the SSTI in black cracking clay with alfalfa sown across a single SSTI line. The SSTI line was 20 cm wide and was installed 30 cm below the surface. The alfalfa germinated directly above the SSTI line but covered 75 cm either side (a total of 1.5 m) in uniform lines. As a result of this research, most SSTI systems are between 6 cm and 20 cm wide. However, some products are up to 2 m wide.

Installation Technologies

In 1997 the first SSTI plow was tested at CSIRO Griffith using a standard three point linkage behind a tractor. This demonstrated that SSTI could be installed at 20–40 cm deep quickly and easily using the plow. SSTI plows are now available to install 1, 2 or 3 lines (more lines are possible).

Other uses of SSTI

In 1996 the first ebb and flow mat technology for potted plants was commercialized based on SSTI technology using drip tapes to control the water delivery. This ebb and flow mat form of SSTI proved very effective in producing potted plants, sprouted wheat and barley for animal production and for research purposes in producing seed varieties without the use of any overhead irrigation. It also provided an effective fertilizer delivery platform.

Components

Well-designed SSTI systems are laid out essentially the same as drip systems but in many cases SSTI can be laid in a "serpentine" pattern vastly reducing the number of take-off connections and potential leaks.

Most drip tubes/tapes used in SSTI are pressure compensating. For example, using a 16mm drip tape, a run of up to 180 m can be achieved from one connection. Longer runs of up to 1,000 m can be achieved using lower flow rates per lineal metre and larger diameter drip tape.

The following components make up a typical SSTI installation:

- Pump or pressurized water source to 100-300kPa (14-43psi).
- Water filters or filtration system from 120 micron with suspended solids less than 30ppm.
- Fertigation injector systems.
- Back flow prevention.
- Pressure regulating valves.
- Main line can be LDPE or PVC.
- Solenoid valves or gate valves to control water flow.
- SSTI system laterals.
- Barbed or Spinlock fittings with stainless steel clips.
- Flushing valves at the end of laterals or combined laterals into a flushing line so that regular flushing can remove suspended solids or bacteria that may build up when using recycled water.

Using drip tapes in SSTI means that there is a wide range of commercial fittings making SSTI very easy to install. Fittings are usually spinlock or ringlock devices securing the drip tape using a barb.

Performance of SSTI

Advantages of SSTI

The advantages of SSTI are:

- SSTI is a "permanent" solution if maintained properly. The components are inert and, given that they are situated below the ground, are not subject to the effects of weather, animals, machinery, vandals or other terrestrial conditions.
- Water savings of 50-75% compared with overhead systems.
- Low pressure requirement (also means lower power requirements).
- Yields can be improved up to four times in certain crops.
- Minimal root intrusion to drippers in the SSTI with a deflective tape on top.
- No emitter blockage due to crusting.
- Minimal effect of evaporation.
- Safe use of recycled or treated water.
- Can use the field (for recreation or agriculture) while irrigation is running.

- Weed growth is minimized because the water does not reach the surface (saving herbicide cost). Germination of weeds only occurs during rainfall.
- Fertigation can be done directly to the root zone (saving fertilizer cost).
- Efficient distribution of nutrients to the entire root zone.
- Broad wetting patterns (moisture covers the entire root zone).
- Water delivery can match the natural capillary rates in soil so saturation is minimized.
- Soil moisture can maintained at field capacity (minimised gravitational losses).
- No surface run off.
- Soil erosion is minimized.
- Fields do not have to be perfectly level.
- Fields with irregular shapes can be accommodated.
- Distance between lines of SSTI is far greater than drip (lower number of solenoids and other components compared to sprinklers and drip).
- SSTI laterals can be ploughed in behind a tractor on large sites (over 10,000 m per day).
- Foliage remains dry (fungal and bacterial leaf disease is minimized).

Disadvantages of SSTI

The disadvantages of SSTI are:

- Initial capital cost is typically more than overhead irrigation.
- Quality installation is critical. If mistakes made, they are difficult to find.
- Installation using correct fittings must be done.
- Regular maintenance is required to ensure long life.
- Automated control and monitoring systems are preferable (subsurface irrigation does not give any visual indicators to show if it is working or not).
- SSTI is usually not UV-treated so it must be kept out of sunlight until it is installed under the surface.
- Temporary overhead watering may be required to establish turf in hot areas.
- Germination of some agricultural crops may require overhead watering if insufficient rainfall.
- SSTI cannot apply fertilizers or herbicides overhead on the surface.
- Rodents may damage the system (although less than drip systems).

LIFT IRRIGATION

Periyar lift irrigation in Kadungallur, Kerala.

Forebay, Stage-1 of Dehani lift irrigation scheme, Maharashtra.

Lift irrigation is a method of irrigation in which water is not transported by natural flow, (as in gravity-fed canal) but is lifted with pumps or surge pools etc.

Lift irrigation schemes must accomplish two main tasks: first, to carry water by means of pumps or other way, from the water source to the main delivery chamber, which is situated at the top most point in the command area. Second. they must distribute this water to the field of the beneficiary farmers by means of a suitable and proper distribution. So that in Lift Irrigation system, the gravity flow of water by canals or river is not available or used.

Of these two issues, distribution is the most complex, for several reasons. First, whereas the system for collecting water at the main delivery chamber makes use of pumps, the distribution system is typically completely gravity-based, and have to be designed solely on the basis of the available hydraulic head. In addition, the area to be irrigated is scattered and has varying topography, and each farm is of a different size, so it can be a challenge to design a gravity-based system that allots each farmer a fair amount of water. Several mechanisms have been developed to meet these challenges.

In the most prevalent system, the area under irrigation is divided into different blocks according to topography determined by contour plan. The water requirement for each block is determined, and pipe lines are designed for these blocks depending on the available head and required discharge. Field delivery chambers are then constructed on the distribution pipe lines, and valves are fitted to control the flow of the pipe lines, with flow from the main delivery chamber controlled by these valves. In this system, lines for lower fields have more discharge, because the force of gravity is greater, so water distribution can be unequal.

In other systems, the area is divided into three or four equal parts, and each part is given the entire discharge for one day. This means each part gets rotation of the full design discharge on the fourth or fifth day. Again, however, water distribution can be unequal, and alternate methods of distribution have been proposed to deal with this problem.

In LI water is lifted from lower level to higher level with the help of pumps and other equipment. Construction of dams and canals helped tremendously to increase the irrigated area lying at lower level than the dam level, but scarcity of water remained the problem for higher level areas. So as to bring higher level area under irrigation L.I.S. are taken up. The 8% of Maharashtra irrigation is occupied by lift irrigation.

Advantages of Lift Irrigation

1. Lift irrigation made irrigation possible at higher level.
2. Land acquisition problem in L.I.S. is less.
3. Water losses are low.
4. Man power is less used.

Requirements of Lift Irrigation System (LIS)

Water Source-Suitable and constant water source should be available for whole year at the site of LIS. Water can be made available through Dams, Weir, River canal etc.

Lifting Medium-To lift water to the desired location, lifting medium i.e. pumps are required. Depending upon the Duty point head, discharge the different types of pumps are selected.

Conveying Medium-Rising main may be of steel, concrete or any other suitable material.

Lift Irrigation Scheme in Telangana

In India, the National Bank for Agriculture and Rural Development reports that there are 1401 lift irrigation schemes financed by them. These schemes irrigate 22000 hectares of land, and have a total financial outlay of Rs. 6462.5 million. However, as many as 573 of these schemes are defunct, mainly due to poor distribution of water. Other schemes have been financed by the government, sugar factories, or individual farmers.

Water shortage is a huge predicament for a country like India which still relies on monsoon rains for much of its agricultural practices. As weather patterns become increasingly uncertain with time, most likely due to global warming, being prepared to cope with water shortages in the future is essential. Rural communities as well as huge population which are greatly affected by water scarcity are the people living in drought-prone areas or in semi-arid up-lands. Much of these disadvantaged areas are also inhibited by tribal villages in many parts of India and are also backward in agricultural practices. These tribal villages are mainly self-sustained in their living habits and rely heavily on the paddy crops for endurance. It is becoming progressively more intricate to maintain tribal lifestyle due to uncertain weather patterns resulting in dwindling of agricultural production over time and all the more increasing the burden on state to raise their standard of living in terms of food security, education, and health care. It is also seen, when time gets tough, villagers generally tend to migrate on a seasonal basis to seek other means of livelihood. Such seasonal migration of entire village population also affects their settled life hampering their children's education as well as their access to other developmental support coming from state.

It is consequently important to help their agriculture instead in order to reduce the incidence of migration in such areas. This can only be achieved by ensuring there is sufficient crop to feed villagers and to sell at market to allow income generation. Irrigation is the essential input to the soil for any vegetation to grow. It is used to lend a hand in the growing of agricultural crops, maintenance of landscapes, and potentials of vegetation by restoring humidity/humus in depleted soils of dry areas and during periods of insufficient rainfall. Additionally, irrigation also has a few other uses in crop production, which include shielding plants against frost, suppressing weed and wild vegetation in grain fields and helping in preventing soil conservation. In contrast, agriculture that relies only on direct rainfall is referred to as rain-fed agriculture, or if it falls in semi-arid regions, it is called dry land farming. Irrigation is often studied together with drainage, which is the natural or artificial removal of surface and sub-surface water from a given area. Various types of irrigation techniques differ in how the water obtained from the source, and is distributed within the field. In general, the aim is to supply water homogeneously to the entire field, so that each land patch has enough amount of water it needs, neither too much nor too little. Irrigation Schemes primarily comprises canal irrigation and lift irrigation schemes. In such schemes, the most important and yet the most neglected part is the local ownership and institutions of farmers taking operative and m responsibilities.

Participatory Irrigation Management

The main purpose of the irrigation reform programme in Andhra Pradesh is to improve the sustainability and productivity of irrigation through sharing irrigation management responsibility with the farmers organized as Water Users Associations. Towards this the State government enacted the Andhra Pradesh Farmers Management of Irrigation System Act in 1997. Presently, water users' organizations under the APFMIS Act is constituted at three different levels in major irrigation projects - Water Users Associations (WUAs) at the minor canal level, Distributor Committees (DCs) at the distributor canal level and Project Committee (PCs) at the main canal level. In medium irrigation project they are constituted at two levels – WUA at the minor canal level and PC at the main canal level. In minor irrigation tanks only a WUA is constituted covering the whole tank command. A total of 10,748 WUAs have been constituted – 2261 in major irrigation projects, 410 in medium irrigation projects and 8077 in minor irrigation tanks.

The water users' organizations undertake the responsibilities of seasonal crop planning and irrigation water regulation, O&M works on canal system and assist in tax collection. To facilitate the functioning of the water users' organizations the State government re-ploughs 100% of the irrigation water tax collected from them. The water tax collection and plough back for O&M to water users' organizations over the last 5 years is given in the table:

	Collection (in $ million)	O&M Plan (in $ millions)	Plough Back (in $ millions)	Budgetary Support (in $ millions)
2005-06	560	NA	0	0
2006-07	670	NA	300	0
2007-08	750	NA	300	0
2008-09	1150	2000	600	1400
2009-10	1320 (Projected)	1600	1000	600

TIDAL IRRIGATION

Tidal irrigation is the subsurface irrigation of levee soils in coastal plains with river water under tidal influence. It is applied in (semi) arid zones at the mouth of a large river estuary or delta where a considerable tidal range (some 2 m) is present.

Technical Description

Tidal heights of 1.3 m and 1.0 m exist in the Gambia River at Jahally and Pacharr, respectively. Special intake structures with gates were constructed, which, when opened at high tide, allow brackish tidal water to enter irrigation channels leading to the farms to be irrigated. The gates are opened for a period of between 3 and 24 hours depending on the size of the area to be irrigated. Tidal irrigation is coordinated with pumped irrigation, and is used to irrigate low lying areas along the Gambia River. Pumped irrigation systems are used for areas at higher elevations.

Operation and Maintenance

Once installed, this technology can be maintained by locally trained artisans with general farm experience.

Level of Involvement

The technology is locally or community-controlled, although some expertise is required for the design, siting, construction and maintenance of the civil structures.

Costs

The capital costs of the project, at 1983/84 prices, was $7.5 million. The current operation and maintenance cost is about $220/ha/year.

Effectiveness of the Technology

The technology has been very successful in paddy rice cultivation. Using tidal irrigation, double cropping of 167 ha and 850 ha is achieved annually at Jahally and Pacharr, respectively. Similarly, irrigation of 440 ha and 125 ha is achieved annually at Jahally and Pacharr, respectively, with pumped irrigation. The annual yield is 9 t/ha/year at an annual cost of $70/t/year.

Suitability

This technology is appropriate in areas where tidal and river conditions are similar to those at Jahally and Pacharr. Generally, the river is in a relatively flat basin with an high degree of tidal intrusion. This technology should only be used with salt-tolerant crops.

Environmental Benefits

There are no known environmental benefits. Drawbacks include a risk of salination and creation of a barrier to migratory fishes.

Advantages

The technology is very successful because, once the intake structures and irrigation channels have been constructed, the operation and maintenance costs are very low. Maintenance work on the irrigation channels, and clearing of weeds and bush from the channels and arable area, can be done by the local farmers.

Disadvantages

The major difficulty experienced is the lack of availability of spare parts.

Cultural Acceptability

It is accepted by the local community and is seen as an income generating project.

SUPPLEMENTAL IRRIGATION

Supplemental irrigation (SI) can be defined as the addition of small amounts of water to essentially rainfed crops during times when rainfall fails to provide sufficient moisture for normal plant growth, in order to improve and stabilize yields.

Supplemental irrigation (SI) has been a promising practice to overcome the constraints outlined above. SI is defined as the addition of limited amounts of water to essentially rainfed crops to improve and stabilize yields when rainfall fails to provide sufficient moisture for normal plant growth. SI is the opposite of conventional irrigation. In the latter, the principal source of moisture is fully controlled irrigation water, and highly variable limited precipitation is only supplementary. SI is dependent on the precipitation as a primary source of water for the crop. SI is an effective response to alleviating the adverse effects of soil moisture stress on the yield of rainfed crops during dry spells. A shortage of soil moisture in rainfed areas often occurs during the most sensitive stages of crop growth (flowering and grain filling). As a result, rainfed crop growth is poor and yield is consequently low. Supplemental irrigation, especially during critical crop growth stages, can improve crop yield and water productivity. Substantial increases in rainfed crop yields, in response to the application of relatively small amounts of water, have been observed (Oweis and Hachum 2003). When rainfall is low, more water is needed, but the response is greater, and yield increases are remarkable even when rainfall is as high as 500 mm.

Benefits of the Practice

- Supplemental irrigation is a simple but highly effective technology that allows farmers to plant and manage crops at the optimal time, without being at the mercy of unpredictable rainfall.
- ICARDA and its partners have developed and optimized supplemental irrigation packages for different crops and cropping systems in Ethiopia, Iran, Jordan, Lebanon, Pakistan, Morocco, Syria, Tunisia and Turkey. Several countries in sub Saharan Africa, including Burkina Faso and Niger, have launched their own programs based on this model. Supplemental irrigation provides multiple benefits: higher and more stable yields, lower risk of crop failure, and significantly higher water productivity (the amount of grain or biomass produced per unit of water).
- Field trials in several countries showed massive increases in wheat and barley yields with small quantities of supplemental irrigation: yields increased from 1.25 to 3 t ha^{-1} in Syria, from 4.6 to 5.8 t ha^{-1} in Morocco, and from 2.2 to 3.4 t ha^{-1} in Iran.
- Supplemental irrigation allows farmers to plant their crops early, increasing yields and preventing exposure to terminal heat and drought stress in hot areas, and frost in cold areas.

Attention is drawn to the importance of sound management of water resources and integrated planning of climate-smart irrigation interventions at all relevant levels from the field scale to the irrigation scheme, river basin and national scales, considering both irrigation 'hardware' (technology, equipment, etc.) as well as irrigation software (policies, institutions, etc). The Compendium on Climate-Smart Irrigation (Batchelor and Schnetzer, forthcoming) provides more details on these issues.

Challenges to Adoption of Supplemental Irrigation Practice

- Soil texture is a major deciding factor when introducing SI. Sandy soils have low water holding capacity and high rates of water infiltration compared to soils with higher levels of clay content. The irrigation system capacity and discharge rate should be equal to or less than the infiltration rate of the soil.
- The crop is another important factor in deciding irrigation depth. Some crops are more water-requiring than others. Knowing how much is growing-season average rainfall, how much is the crop water requirement and how much is the deficit that needs to be met using SI is important information when planning water resources availability for an SI project.
- Landscape of the irrigation site is another important criterion. If the land is uneven, water cannot flow at a constant rate and cannot reach every corner of the field. In such situations, sprinklers or drips are recommended. Both these methods are relatively expensive and need energy to operate. If the crop is predominantly rainfed and only needs to be irrigated a few times during the growing season, low installation and maintenance costs are a major factor for adoption by the farmer. Solid set or moving sprinkler systems are cheaper to use than drip irrigation systems.
- Capacity of the reservoir should be such that it can meet the demand of the crop. It is very expensive to excavate, especially in remote locations. When deciding on the capacity, it should be sufficient to meet crop water demands and, in the case of water harvesting, the runoff generated from the upstream catchment area. If the catchment area is insufficient, the reservoir will not get filled and will be unable to meet the crop water demand.
- Supplemental irrigation depth and timing is an important decision. SI is not intended to meet the full crop water demand. It is a critical dose which can increase yields significantly as well as save the crop from failure during a dry year. The depth of SI needs to be fixed when designing the water storage reservoir and the irrigation system. It is better to apply small doses of SI rather than all in 1–2 operations. Smaller doses give the crop an opportunity to use all the water and not let any go waste as deep percolation or runoff. However, there are additional costs involved in applying more doses. So, a balance needs to be achieved between how often and how much water you apply in each dose. But it is not easy to perform such calculations and make decisions without sufficient knowledge of the practice. Capacity building of the farmers is a challenge needed to overcome this obstacle.

Where can Supplemental Irrigation be Practiced?

In the lowlands of the Mediterranean region and rainfed south Asian countries, farmers usually sow/seed their land only when a sufficient amount of rain has fallen. The date by which this amount of rain has fallen is usually called the 'onset rainfall', meaning the beginning of the rainy season. Here continuous cropping prevails as pure cereal or cereal-legume rotations. Every week's delay after this time results in a yield decrease of between 200 and 250 kg/ha. If the onset of the seasonal rain is delayed, early sowing can be realized with the help of a SI system. With SI it is possible to decide on the sowing date of the basically rain-fed crops without needing to wait for the onset of the seasonal rain. This results in a longer growing season, better yield, and an earlier maturity that helps crops to escape terminal drought. In the highlands, frost conditions occur between December and March and field crops remain dormant. Usually, the first rainfall, sufficient to germinate the seeds (the onset rain), comes late and results in a small crop stand when the frost occurs.

As a result, rainfed yields are much lower than when the crop stand, pre-frost, is good. Ensuring a good crop stand before frost can be achieved by early sowing and applying from 50 to 70 mm of SI. Supplemental irrigation, given at early sowing, dramatically increases wheat yield and water productivity. In the highlands of Turkey, applying 50 mm of SI to wheat sown early has increased grain yield by more than 60%, adding more than 2 t/ha to the average rain-fed yield of 3.2 t/ha. Water productivity (WP) reached 4.4 kg/m^3 of consumed water compared to WP values for wheat from 1 to 2 kg/m^3 under traditional practices. Similar results were found in the Iran highlands for wheat and barley.

Supplemental Irrigation to Alleviate Moisture Stress

Unlike for full irrigation, the timing for SI cannot be determined in advance. This is because the basic source of water to rainfed crops is rainfall which, being variable in amount and distribution, is difficult to predict. Since SI water is best given when the soil moisture drops to a critical level, the time for irrigation can be best determined by measuring the soil moisture on a regular basis. Supplemental irrigation is even referred to as life-saving irrigation in some south Asian countries. Usually, however, one to three supplemental irrigation applications of not more than 100 mm each annually appear sufficient, depending on the rainfall amount and distribution.

Contribution to CSA Pillars

1. How does supplemental irrigation practice increase productivity, farm livelihoods and food security?

The area of wheat under SI in northern and western Syria (where annual rainfall is greater than 300 mm) increased from 74,000 ha (in 1980) to 418,000 (in 2000), an increase of 470%. Estimated mean annual income in production cost due to SI (including fixed and variable costs) as compared with rainfed equals USD 150 per hectare. Estimated mean increase in net profit between rainfed and SI for wheat equals USD 300 per hectare. The ratio of increase in estimated annual net profit per hectare to estimated difference in annual costs between rainfed and SI is 200%, which is high. But increase in crop production per unit of land or per unit of water does not necessarily increase farm profit, just because of the nonlinearity of crop yield with production inputs, particularly with water and its interaction with other input factors (such as fertilizer, sowing date, choice of cultivar, irrigation system, etc). Therefore, a water management strategy that maximizes yield or water productivity is not necessarily the most desirable one, especially in water-scarce areas. Often such a strategy is not the most economical in terms of net return. Actually, the most desirable strategy is somewhere in between these two.

2. How does supplemental irrigation help adapt to and increase resilience to climate change impacts?

Climate change adaptation strategies can only be effective if done in an integrated manner. For rainfed agriculture, the strategies may need to encompass water management, crop improvement, cultural practices, policies, and socioeconomic and other issues. Supplemental irrigation, however, can play an important role in the adaptation efforts to climate change in rainfed agroecosystems. Following are some of the potential responses that SI can make to adapt to climate change:

- ICARDA is currently investigating climate change effects using biophysical crop models. First scenario simulation results highlight the positive effect of SI in alleviating the negative impact of climate change, especially on the year-to-year variability in crop yields which are predicted as consequences of climate change in the near future. Supplemental irrigation management resulted in an application of irrigation water of between 0 and 330 mm (average, 122 mm), usually, between one and three (maximum, six) irrigation events per season. No supplemental irrigation was necessary in six of the 49 years. Thus, on average, yields were 2.6 times higher with SI than without. The corresponding wateruse efficiency (precipitation plus irrigation) increased from 0.50 kg/m^3 under rainfedonly conditions to 0.99 kg/m^3 under SI (applied to meet 100% of crop water requirement). Therefore, in conclusion, it can be noted that SI can be used to overcome the changes in soil-water-plant relations, especially in alleviating soil water stress resulting from changes in crop ET and crop patterns. As rainfall is unpredictable, SI becomes the most viable practice to alleviate the moisture stress caused by increased temperature. Another adaptation option is the possibility of changing planting dates. With SI, this can also help adaptation to global warming. With the help of SI, early planting is possible and the growing season can start relatively early.

- Less and more erratic precipitation is expected in the dry areas as a result of global warming. Lower precipitation will cause a further moisture stress on already stressed rainfed crops and some areas on the peripheries of the rainfed zones maydrop out of dryland agriculture as a result. It is also expected that rainfall will be more erratic and intensive, and the season will have prolonged drought spells. Crop yields and WP losses are mainly associated with soil moisture stress during such drought spells. Prolonged drought spells during the rainy seasons resulting from global warming will make the crop situation even worse and further drops in yields are expected as a result. Supplemental irrigation, by definition, deals with two situations. It adds some water to compensate for lower rainfall and less moisture storage and it alleviates soil water stress during dry spells. It is however, important to quantify the changes in rainfall characteristics and the durations of potential drought spells in order to design SI schedules to adapt the system to climate change.

- Higher intensity rainstorms are also predicted, not only in the dry area, but also in SSA and globally. This naturally will cause more runoff and soil erosion in rainfed areas, especially on sloping lands. The portion of the precipitation that normally infiltrates the soil to support plants growth will be less as more runoff will go downstream. In addition, it is predicted that the process of land degradation will be accelerated. Supplemental irrigation combined with water harvesting can provide workable solutions to this problem. Macro- and microcatchment water harvesting are effective strategies for intercepting runoff and storing water either in the soil profile or in surface and groundwater aquifers. Water stored in the soil may support plants directly or it can be used for SI during dry spells if stored in small reservoirs or ground water aquifers. This model is being researched and tested in many places and should provide a good platform for overcoming the effects of climate change on runoff.

3. How does supplemental irrigation mitigate GHG emission?

Supplemental irrigation can help mitigate greenhouse gas (GHG) emissions in two ways:

- Increased yields achieved with SI (compared to rainfed production) result in higher carbon sequestration rates in plant biomass and for the build-up of soil organic carbon (SOC), especially if crop residues are returned to the soil or if cover cropping practices keep the soil protected as much as possible. This increase in SOC, in turn, has a synergistic effect on adaption through improved soil fertility as well as reduced nutrient leaching and soil erosion. If this soil fertility improvement results in a decrease of mineral fertilizer use, additional mitigation benefits can be obtained through a reduction of GHG emissions linked to fertilizer manufacturing and use.

- When substantial yield gains are achieved through SI compared to a relatively low increase in farming inputs, the GHG emission intensity per unit of produce decreases. Although this may imply higher GHG emissions per area, the yield gains can prevent conversion of additional areas to crop land and associated release of carbon from soils. The GHG emission intensity can be further reduced by the use of renewable energies instead of fossil fuels for SI, e.g. through solar-powered irrigation systems that lower the carbon dioxide emissions associated to the delivery of irrigation water. Also the production of biofuel crops using SI that displaces the use of fossil fuels on farm or in the wider economy can contribute to mitigate GHG emissions. Biofuel crops should be considered only in contexts and at scales that do not negatively affect the food security of the local population through competition with food crops.

Costs and Funding for Supplemental Irrigation

Risk weighs on the daily lives of poor rainfed farmers, and investment packages have to help reduce that risk. Risks include not only climatic and limited access to reliable technology and water, but also unstable land tenure and poorly functioning product and credit markets. Investing in SI can have a significant impact on justified returns from dry farming systems. Biophysical returns on water with SI are higher than those under conventional irrigation, and are highest with deficit applications. An integrated investment package including waterharnessing and irrigation technology, irrigation scheduling, training, and cropping and fertilizing guidance is probably the best. Combined soil and water management investments can also have a high return. The key requirements for successful investment in SI include:

- Determining the most appropriate scheduling, crops and cropping patterns, and socioeconomic feasibility.

- Strong water-user associations with incentives for local communities to use water efficiently.

- Managing the economic and environmental consequences of using water in SI.

- Developing policies that foster an enabling environment for the adoption of waterefficient technologies.

Metrics for CSA Performance of Supplemental Irrigation

The effect of supplemental irrigation goes beyond yield increases to substantially improving the water productivity. Both the productivity of irrigation water and that of rainwater are improved when they are used conjunctively. Water productivity in supplemental irrigation is a function of the amount of irrigation water applied. It has been found that maximum water productivity is attained when from one- to two-thirds of the full irrigation water is applied.

Interaction with other CSA Practices

Supplemental irrigation alone, although it alleviates moisture stress, cannot ensure the best performance of the rain-fed agricultural system. It has to be combined with other farm management practices and inputs such as land preparation, fertility and pest management, and choice of right crop varieties for the location and climate. According to Rockström et al., supplemental irrigation is a key strategy, still underused, for unlocking rainfed yield potential and water productivity.

Supplemental Irrigation in the Ghanaian Farming System

Rainfed agriculture provides about 95 percent of Ghanaian food and feed, and it is the main source of livelihood for more than 70 percent of the population. Therefore, any inadequate rainfall (intensity or timing) at key periods during the crop growth cycle (seeding, growing, flowering and grain filling) affects crop yields and farmers income which in turn may pull framers into poverty trap. Rainfall is the most variable climatic parameter and as such shows wide variability differences over ecological regions. In Ghana, based on 1961 to 2000 climate statistics, annual rainfall amounts are expected to decrease by 1.1 to 3.1 percent across all ecological regions by the year 2020 and by 13.0 to 21.0 percent by 2080. SI is an effective response to alleviating these adverse effects of rainfall shortage or dry spell on the yield of rainfed crops. It of supplying crops with water during key stages of crop growth cycle when rainfall fails to provide sufficient water.

SI is an old practice in Ghana. It started with the involvement of the Ghanaian government in agriculture in the 80s through construction of runoff reservoirs that were then used by farmers for both wet and dry seasons farming. SI in Ghana represents 0.8 percent of the country entire irrigation land use mostly based in the Northern Savannah agro-ecological zone (Northern, Upper East and Upper West) which is the largest agriculture zone in Ghana.

Through a three years USAID project on drought risk management in Northern Savannah agro-ecological zone, which one key objective was to access supplemental irrigation use and demand, it was found that nowadays many farmers in drought prone area through their association or community based organization are able to dug micro ponds to collect and store runoff water from neighbouring farm micro-watersheds. The pond is man-made and is dug in an upstream location from the irrigated plot to collect runoff. Adjustments are made in the ground to facilitate runoff pouring into the pond. The pond water is then applied to rainfed crops using watering cans and in some areas moto pumps using hose to apply to the crops. Some farmers own seasonal shallow wells which range in depth from 1 to 5 meters. A well can be dug by one person in approximately 6 hours at a cost of around USD 5/hour and it can serve plots within a 50 to 60 meters radius.

Seasonal shallow well located at Kupulima community Ghana.

Water is collected from the wells using ropes and buckets. Seasonal shallow wells are commonly used in low-lying areas with high water tables and they are for wet season farming as they are mostly dried in dry season.

- Farmers under the government irrigation facilities pay a user fee of GHS 20 (USD 5) per acre of land to have access to water during wet season. While under the community-based irrigation facilities farmers are to pay a maintenance fees of GHS 4 (USD 1). However, through a contingent valuation assessment, it was found that the average Willingness-to-Pay (WTP) for SI per acre of land is GHS 26 (USD 6.5) in the Northern Savannah agro-ecological zone of Ghana.

- The benefits of the SI system, however, are more evident in the Northern Savannah agro-ecological zone of Ghana which is characterized by unimodal rainfall of short duration compared to the rest of the country which is characterized by two rainy seasons. Farmers mostly apply SI in their rice and maize farms. SI provides multiple benefits to farmers: it provides higher and more stable yields, reduces the risk of crop failure, significantly increases water productivity, and has a positive spill-over effect in modern technologies adoption (seed and fertilizer). The gain in maize yield due to SI in the scheme is 2.22 times higher than the average rainfed yield in the Northern Savannah agro-ecological zone. Farmers under SI invest 20 percent more in farm inputs per hectare compared to the average per hectare farm inputs investment in Northern Savannah agro-ecological zone. Through this investment farmers under SI are able to increase their per hectare income by 35 up to 50 percent.

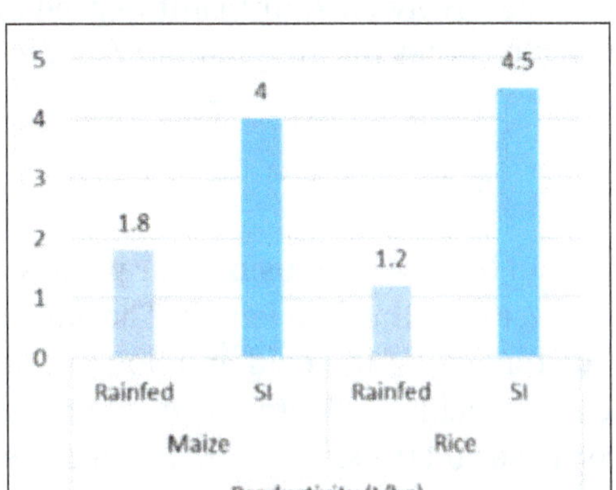

Maize and rice productivity under supplemental irrigation (SI) and rainfed production in Northern Savannah agro-ecological zone of Ghana.

- Though SI is an old practice in Ghana, few farmers have adopted it. The main challenges of its adoption include: lack of water resources, lack of knowledge transfer, landownership, high-lying lands, and poor farmers often cannot afford the construction, maintenance and operational costs given their limited resources. There is a need for the government and other development agencies to provide incentives and facilities through which farmers may be able to adopt supplemental irrigation. Construction of large and small scale runoff reservoirs, farmers' capacity building on the use and maintenance of water resources are some key incentives. Farmers are also to operate in cooperative or farmers based organization to gain economy of scale as natural resources, particularly water, are more efficiently utilized on a collective basis than individually.

SURFACE IRRIGATION

Furrow Irrigation of sugar cane.

Surface irrigation is where water is applied and distributed over the soil surface by gravity. It is by far the most common form of irrigation throughout the world and has been practiced in many areas virtually unchanged for thousands of years.

Surface irrigation is often referred to as flood irrigation, implying that the water distribution is uncontrolled and therefore, inherently inefficient. In reality, some of the irrigation practices grouped under this name involve a significant degree of management (for example surge irrigation). Surface irrigation comes in three major types; level basin, furrow and border strip.

Process

The process of surface irrigation can be described using four phases. As water is applied to the top end of the field it will flow or advance over the field length. The advance phase refers to that length of time as water is applied to the top end of the field and flows or advances over the field length. After the water reaches the end of the field it will either run-off or start to pond. The period of time between the end of the advance phase and the shut-off of the inflow is termed the wetting, ponding or storage phase. As the inflow ceases the water will continue to runoff and infiltrate until the entire field is drained. The depletion phase is that short period of time after cut-off when the length of the field is still submerged. The recession phase describes the time period while the water front is retreating towards the downstream end of the field. The depth of water applied to any point in the field is a function of the opportunity time, the length of time for which water is present on the soil surface.

Types of Surface Irrigation

Basin Irrigation

Level basin irrigation has historically been used in small areas having level surfaces that are surrounded by earth banks. The water is applied rapidly to the entire basin and is allowed to infiltrate. In traditional basins no water is permitted to drain from the field once it is irrigated. Basin irrigation is favoured in soils with relatively low infiltration rates. This is also a method of surface irrigation. Fields are typically set up to follow the natural contours of the land but the introduction of laser levelling and land grading has permitted the construction of large rectangular basins that are more appropriate for mechanised broadacre cropping.

Level basin flood irrigation on wheat.

Residential flood irrigation.

Drainback Level Basins

Drain back level basins (DBLB) or contour basins are a variant of basin irrigation where the field is divided into a number of terraced rectangular bays which are graded level or have no significant slope. Water is applied to the first bay (usually the highest in elevation) and when the desired depth is applied water is permitted to drain back off that bay and flow to the next bay which is at a lower elevation than the first. Each bay is irrigated in turn using a combination of drainage water from the previous bay and continuing inflow from the supply channel. Successful operation of these systems is reliant on a sufficient elevation drop between successive bays. These systems are commonly used in Australia where rice and wheat are grown in rotation.

Furrow Irrigation

Furrow irrigation is conducted by creating small parallel channels along the field length in the direction of predominant slope. Water is applied to the top end of each furrow and flows down the field under the influence of gravity. Water may be supplied using gated pipe, siphon and head ditch, or bankless systems. The speed of water movement is determined by many factors such as slope, surface roughness and furrow shape but most importantly by the inflow rate and soil infiltration rate. The spacing between adjacent furrows is governed by the crop species, common spacings typically range from 0.75 to 2 metres. The crop is planted on the ridge between furrows which may contain a single row of plants or several rows in the case of a bed type system. Furrows may range anywhere from less than 100 m to 2000 m long depending on the soil type, location and crop type. Shorter furrows are commonly associated with higher uniformity of application but result in increasing potential for runoff losses. Furrow irrigation is particularly suited to broad-acre row crops such as cotton, maize and sugar cane. It is also practiced in various horticultural industries such as citrus, stone fruit and tomatoes.

Furrow irrigation system using siphon tubes.

Gated pipe supply system.

The water can take a considerable period of time to reach the other end, meaning water has been infiltrating for a longer period of time at the top end of the field. This results in poor uniformity with high application at the top end with lower application at the bottom end. In most cases the performance of furrow irrigation can be improved through increasing the speed at which water moves along the field (the advance rate). This can be achieved through increasing flow rates or through the practice of surge irrigation. Increasing the advance rate not only improves the uniformity but also reduces the total volume of water required to complete the irrigation.

Surge Irrigation

Surge irrigation is a variant of furrow irrigation where the water supply is pulsed on and off in planned time periods (e.g. on for 1 hour off for 1½ hour). The wetting and drying cycles reduce infiltration rates resulting in faster advance rates and higher uniformities than continuous flow. The reduction in infiltration is a result of surface consolidation, filling of cracks and micro pores and the disintegration of soil particles during rapid wetting and consequent surface sealing during each drying phase. On those soils where surging is effective it has been reported to allow completion of the irrigation with a lower overall water usage and therefore higher efficiency and potentially offer the ability to practice deficit irrigation. The effectiveness of surge irrigation is soil type dependent; for example, many clay soils experience a rapid sealing behaviour under continuous flow and therefore surge irrigation offers little benefit.

Bay/Border Strip Irrigation

Border strip, otherwise known as border check or bay irrigation could be considered as a hybrid of level basin and furrow irrigation. The field is divided into a number of bays or strips, each bay is separated by raised earth check banks (borders). The bays are typically longer and narrower compared to basin irrigation and are orientated to align lengthwise with the slope of the field. Typical bay dimensions are between 10-70m wide and 100-700m long. The water is applied to the top end of the bay, which is usually constructed to facilitate free-flowing conditions at the downstream end. One common use of this technique includes the irrigation of pasture for dairy production.

Issues Associated with Surface Irrigation

While surface irrigation can be practiced effectively using the correct management under the right conditions, it is often associated with a number of issues undermining productivity and environmental sustainability:

- Waterlogging - Can cause the plant to shut down delaying further growth until sufficient water drains from the rootzone. Waterlogging may be counteracted by drainage, tile drainage or watertable control by another form of subsurface drainage.

- Deep drainage - Overirrigation may cause water to move below the root zone resulting in rising water tables. In regions with naturally occurring saline soil layers (for example salinity in south eastern Australia) or saline aqifers, these rising water tables may bring salt up into the root zone leading to problems of irrigation salinity.
- Salinization - Depending on water quality irrigation water may add significant volumes of salt to the soil profile. While this is a lesser issue for surface irrigation compared to other irrigation methods (due to the comparatively high leaching fraction), lack of subsurface drainage may restrict the leaching of salts from the soil. This can be remedied by drainage and soil salinity control through flushing.

The aim of modern surface irrigation management is to minimse the risk of these potential adverse impacts.

GREEN LAWN IRRIGATION

Green lawn irrigation not only saves water and helps the environment; it results in a healthier landscape. Turf management studies have shown that most lawns only need to be watered every 4 to 8 days, yet many homeowners irrigate much more frequently.

One of the key concepts in green lawn irrigation is that plants that are watered deeply create deeper, healthier root systems. By reducing the number of lawn irrigation days, and increasing the duration of watering, a healthy, stable lawn can be created.

One or two good sessions a week should provide enough water for the lawn, but runoff should be watched as the duration of irrigation is increased. If puddles or water draining off the lawn is noticed, watering time should be broken into shorter cycles with a 15 or 30-minute break in between.

A regular schedule of fertilizer will give grass the energy to develop into strong, healthy turf, reducing lawn irrigation needs. Fertilize cool season grass in the spring and fall. Fertilize warm season grass in the early and late summer.

Make sure the blade of the mower is sharp and cut grass a bit longer. Long grass blades promote deep roots and provide shade to reduce heat stress. Mowing early in the day reduces the stress created by mowing.

Inspect the lawn irrigation system regularly. Replace broken or leaky sprinkler heads. Check for tilted or misaligned heads and adjust as needed. Clogged heads will reduce the lawn irrigation coverage area and wastewater. Inspect and clean sprinkler heads to remove accumulated grit or plant debris.

Lawn irrigation systems that are run under too much pressure produce mist instead of droplets. Mist is carried away by the wind and never reaches your lawn.

In warm weather, adjust the lawn irrigation system to only water at night. Water loss will be less due to evaporation by avoiding daytime watering.

Override the lawn irrigation in windy or rainy weather. Wind will cause uneven watering and rain is free water-don't duplicate the effort.

LOCALIZED IRRIGATION

Localized irrigation systems apply water directly where the plant is growing thus minimizing water loss through evaporation from the soil. Such localized irrigation systems include drip irrigation, porous clay pots, porous pipes, and perforated plastic sleeves.

Porous Clay Pots

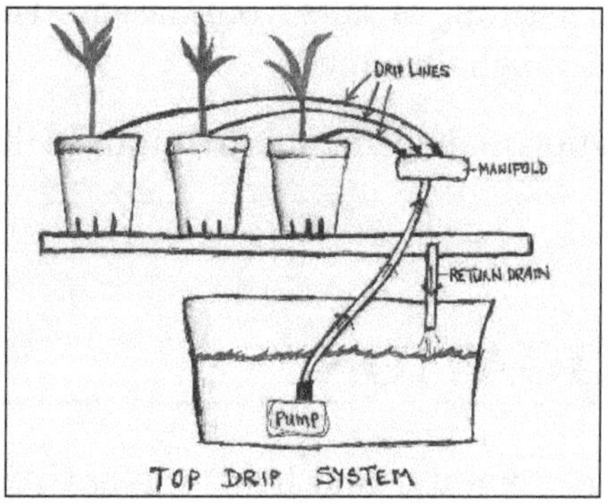

The Pattern of Soil Wetting Around a Porous Clay Pot.

This is a method of irrigation in which water is stored in clay pots buried in the ground, from where it is slowly released to the plants. This method is good for fruit trees. Such use of soil-embedded porous jars is one of the oldest continuous irrigation methods that probably originated in the Far East and North Africa.

The method consists of:

- Clay pots that are placed in shallow pits dug for this purpose;
- Soil is then packed around the neck of the pots so that the necks protrude a few centimetres above the ground surface;
- Water is poured into the pots, either by hand or by means of a flexible hose connected to a water source.

The pots are made of locally available clay with optimum properties of strength (to resist crushing), permeability (to exude water into the soil at an approximately steady rate), and size (to hold enough water for at least one day's supply).

The potential of clay-pot irrigation has not been fully exploited by farmers in the eastern and southern Africa region, even though the technology is suitable for small scale farmers. There have not been many reports of previous experience in the region.

Porous Clay Pipes

Water is spread along a continuous horizontal band in the soil. This method is most suited to closely spaced crops such as vegetables. The locally made clay pipes are approximately 24 cm in length and 7.5 cm in internal diameter, with wall thickness of 2 cm. The pipes are placed at the bottom of a shallow trench (about 25 cm deep) representing the centre line of a 1-m wide bed.

Perforated Plastic Sleeves

Plastic sheeting has been used to make a sleeve-like casing. The advantage of this is the low cost. However, the method has several distinct disadvantages that restrict the range of its applicability. Since the soft plastic material used for making the sleeve does not retain its shape, the sleeve must be filled with sand before being placed in the soil. The sand filling reduces the capacity of the sleeve by some 50–60%. Moreover, the sand itself tends to retain a significant fraction of the moisture and thus to restrict outflow.

Since the plastic casing is essentially impervious, it must be perforated. The difficulty of standardizing the diameter and density of the perforations introduces another variable unto the system. The best configuration must be established by trial and error. This method has been used with success in Senegal. An interesting variation in Sri Lanka consists of a 50-cm long PVC pipe of standard °-inch diameter. The pipe ends in an emitter, a block made from a 1:10 cement/ sand mixture. The design was found to be effective in enhancing the survival and growth rate of young fruit trees.

FURROW IRRIGATION

Furrows are small, parallel channels, made to carry water in order to irrigate the crop. The crop is usually grown on the ridges between the furrows.

When to use Furrow Irrigation

Furrow irrigation is suitable for a wide range of soil types, crops and land slopes.

Suitable Crops

Furrow irrigation is suitable for many crops, especially row crops. Crops that would be damaged if water covered their stem or crown should be irrigated by furrows.

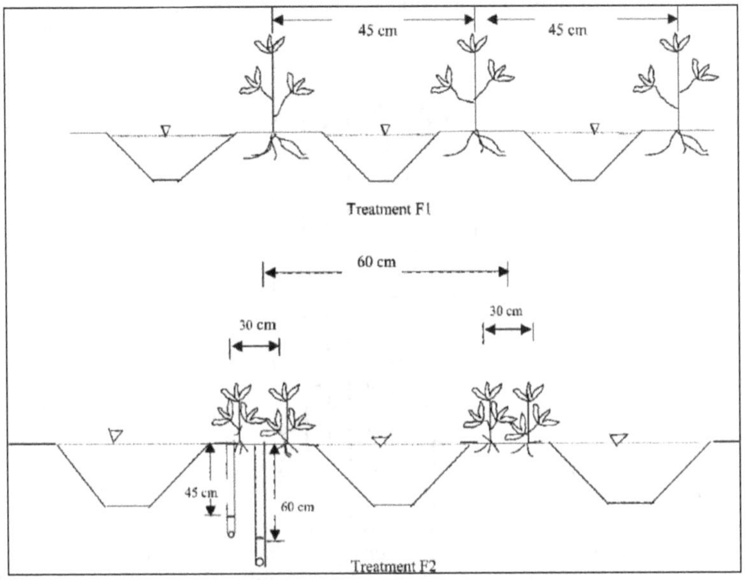

Furrow irrigation is also suited to the growing of tree crops. In the early stages of tree planting, one furrow alongside the tree row may be sufficient but as the trees develop then two or more furrows can be constructed to provide sufficient water. Sometimes a special zig-zag system is used to improve the spread of water.

Corrugation irrigation, frequently mentioned in literature, is a special type of furrow irrigation, used for broadcast crops. Corrugations are small hills pressed into the soil surface. The application of this method is limited and is not included in this manual.

The following crops can be Irrigated by furrow irrigation:

- Row crops such as maize, sunflower, sugarcane, soybean;
- Crops that would be damaged by inundation, such as tomatoes, vegetables, potatoes, beans;
- Fruit trees such as citrus, grape;
- Broadcast crops (corrugation method) such as wheat.

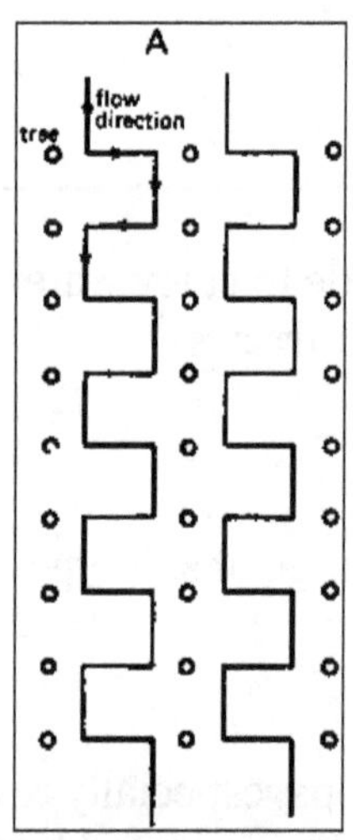

Zig-zag furrows - A: Zig-zag furrows used for irrigating trees on land with a moderate slope (0.5-1.5%).

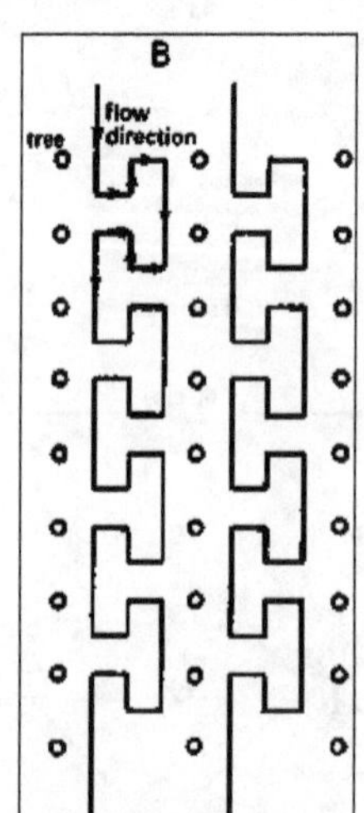

Zig-zag furrows - B: Another zig-zag pattern for furrow irrigation on fairly flat slopes (under 0.5%).

Suitable Slopes

Uniform flat or gentle slopes are preferred for furrow irrigation. These should not exceed 0.5%. Usually a gentle furrow slope is provided up to 0.05% to assist drainage following irrigation or excessive rainfall with high intensity.

On undulating land furrows should follow the land contours. However, this can be a difficult operation requiring very careful setting out of the contours before cutting the furrows.

Suitable Soils

Furrows can be used on most soil types. However, as with all surface irrigation methods, very coarse sands are not recommended as percolation losses can be high. Soils that crust easily are especially suited to furrow irrigation because the water does not flow over the ridge, and so the soil in which the plants grow remains friable.

Furrow Layout

This section deals with the shape, length and spacing of furrows. Generally, the shape, length and spacing are determined by the natural circumstances, i.e. slope, soil type and available stream size. However, other factors may influence the design of a furrow system, such as the irrigation depth, farming practice and the field length.

Furrow Length

Furrows must be on consonance with the slope, the soil type, the stream size, the irrigation depth, the cultivation practice and the field length. The impact of these factors on the furrow length is discussed below:

1. Slope:

Although furrows can be longer when the land slope is steeper, the maximum recommended furrow slope is 0.5% to avoid soil erosion. Furrows can also be level and are thus very similar to long narrow basins. However a minimum grade of 0.05% is recommended so that effective drainage can occur following irrigation or excessive rainfall. If the land slope is steeper than 0.5% then furrows can be set at an angle to the main slope or even along the contour to keep furrow slopes within the recommended limits. Furrows can be set in this way when the main land slope does not exceed 3%. Beyond this there is a major risk of soil erosion following a breach in the furrow system. On steep land, terraces can also be constructed and furrows cultivated along the terraces.

2. Soil Type:

In sandy soils water infiltrates rapidly. Furrows should be short (less than 110 a), so that water will reach the downstream end without excessive percolation losses.

In clay soils, the infiltration rate is much lower than in sandy soils. Furrows can be much longer on clayey than on sandy soils.

3. Stream Size:

Normally stream sizes up to 0.5 l/sec will provide an adequate irrigation provided the furrows are not too long. When larger stream sizes are available, water will move rapidly down the furrows and so generally furrows can be longer. The maximum stream size that will not cause erosion will obviously depend on the furrow slope; in any case, it is advised not to use stream sizes larger than 3.0 l/sec.

4. Irrigation Depth:

Applying larger irrigation depths usually means that furrows can be longer as there is more time available for water to flow down the furrows and infiltrate.

5. Cultivation Practice:

When the farming is mechanized, furrows should be made as long as possible to facilitate the work. Short furrows require a lot of attention as the flow must be changed frequently from one furrow to the next. However, short furrows can usually be irrigated more efficiently than long ones as it is much easier to keep the percolation losses low.

6. Field Length:

It may be more practical to make the furrow length equal to the length of the field, instead of the ideal length, when this would result In a small piece of land left over. Equally the length of field may be much less than the maximum furrow length. This is not usually a problem and furrow lengths are made to fit the field boundaries.

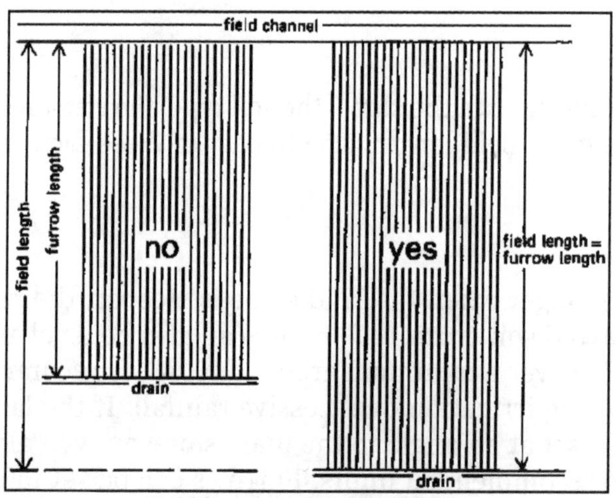

Field length and furrow length.

Table below gives some practical values of maximum furrow lengths under small-scale irrigation conditions. The values shown in table below. These higher values are appropriate under larger scale, fully mechanized conditions.

Table: Practical Values of Maximum Furrow Lengths (M) Depending on Slope, Soil Type, Stream Size and Net Irrigation Depth.

Furrow slope (%)	Maximum stream size (l/s) per furrow	Clay		Loam		Sand	
		Net irrigation depth (mm)					
		50	75	50	75	50	75
0.0	3.0	100	150	60	90	30	45

0.1	3.0	120	170	90	125	45	60
0.2	2.5	130	180	110	150	60	95
0.3	2.0	150	200	130	170	75	110
0.5	1.2	150	200	130	170	75	110

This table only provides approximate Information relating furrow slope, soil type, stream size and irrigation depth to furrow lengths. This should only be used as a guide as the data are based primarily on field experience and not on any scientific relationships. Maximum values of furrow length are given for reasonably efficient irrigation. However, furrow lengths can be even shorter than those given in the table and in general this will help to improve irrigation efficiency. Only by Installing a furrow system, following the guidelines, and then evaluating its performance can an appropriate system be developed for a given locality.

Furrow Shape

The shape of furrows is influenced by the soil type and the stream size.

Soil Type:

In sandy soils, water moves faster vertically than sideways (= lateral). Narrow, deep V-shaped furrows are desirable to reduce the soil area through which water percolates. However, sandy soils are less stable, and tend to collapse, which may reduce the irrigation efficiency.

In clay soils, there is much more lateral movement of water and the infiltration rate is much less than for sandy soils. Thus a wide, shallow furrow is desirable to obtain a large wetted area to encourage infiltration.

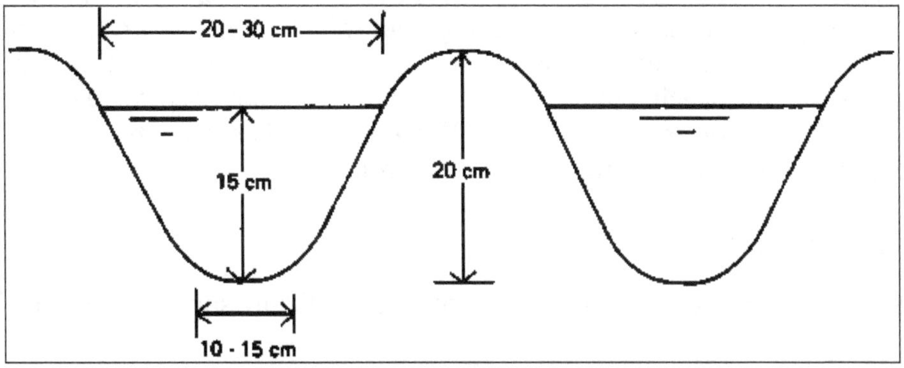

A deep, narrow furrow on a sandy soil.

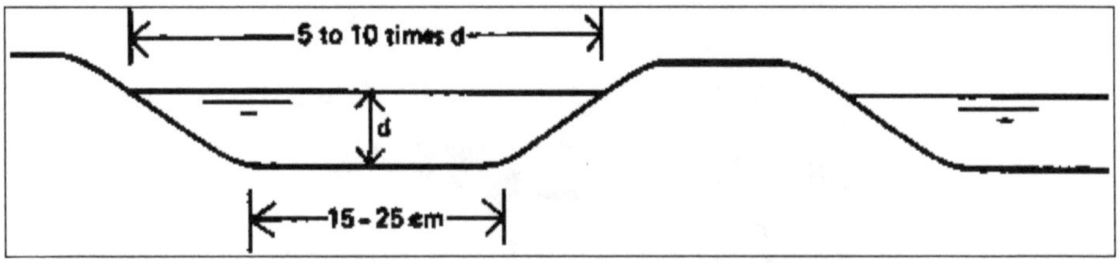

A wide, shallow furrow on a clay soil.

Stream Size:

In general, the larger the stream size the larger the furrow must be to contain the flow.

Furrow Spacing

The spacing of furrows is influenced by the soil type and the cultivation practice.

Soil Type:

As a rule, for sandy soils the spacing should be between 30 and 60 cm, i.e. 30 cm for coarse sand and 60 cm for fine sand.

On clay soils, the spacing between two adjacent furrows should be 75-150 cm. On clay soils, double-ridged furrows, sometimes called beds can also be used. Their advantage is that more plant rows are possible on each ridge, facilitating manual weeding. The ridge can be slightly rounded at the top to drain off water that would otherwise tend to pond on the ridge surface during heavy rainfall.

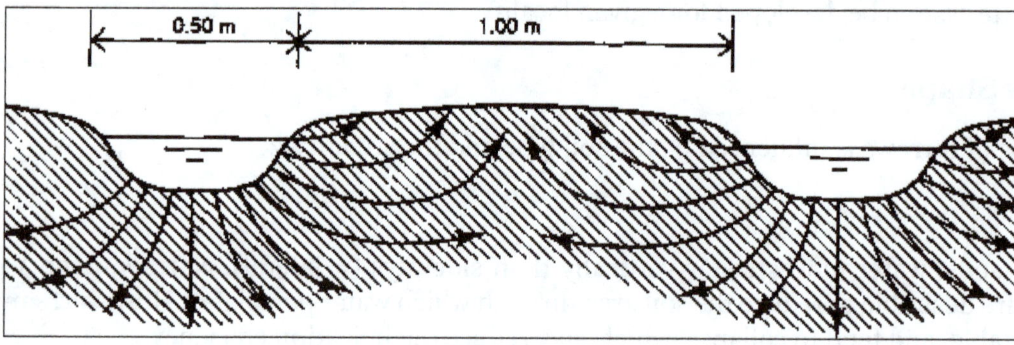

A double-ridged furrow.

Cultivation Practice:

In mechanized farming a compromise is required between the machinery available to cut furrows and the ideal spacings for crops. Mechanical equipment will result in less work if a standard width between the furrows is maintained, even when the crops grown normally require a different planting distance. This way the spacing of the tool attachment does not need to be changed when the equipment is moved from one crop to another. However, care is needed to ensure that the standard spacings provide adequate lateral wetting on all soil types.

Furrow Construction

The most common way to construct furrows is with a ridger. Figure shows animal and hand-drawn ridgers.

Ridger plough: wooden body, animal-drawn.

Ridger plough: iron type, animal-drawn.

Ridger plough: hand-drawn version.

Construction of Furrows on Flat or Mildly Sloping Land

The following steps are taken to construct furrows: setting out; forming one (or more) ridge; forming one (or more) parallel ridge:

Step 1: A straight line is set out in the field along the proposed line of furrows. This can be done by setting up ranging poles or marking a line on the ground with chalk powder or small mounds of earth. An experienced ploughman should be able to plough along the line by aligning the poles or earth mounds by eye.

Step 2: The ridger is moved along the line. The resulting furrow should be straight. If not, the area should be ploughed again and the procedure repeated.

Step 3: About every five metres, a new straight line should be set out.

If a ridger-drawbar connected with a tractor is used, four furrows can be drawn simultaneously. On the track back the left ridger is put in the last furrow track to make sure the new furrows arc parallel to the previous ones. Also here it should be checked that straight lines are followed: for every track a centre line is set out.

Construction of Furrows on Sloping or Undulating Land

Special care is needed to construct furrows along the contour on sloping or undulating land. The following steps are taken to construct furrows along the contour:

Step 1: A guide furrow must first be set out along the upper edge of the field close to the farm channel using a levelling device to locate the contour line. Further guide furrows are set out every 5 metres on undulating ground and every 10 metres on uniformly sloping land.

Step 2: Working from each guide furrow, furrows are made to halfway along the next guide furrow.

Irrigating Furrows

Water is supplied to each furrow from the field canal, using siphons or spiles. Sometimes, instead of the field canal with siphons or spiles, a gated pipe is used.

Depending on the available flow in the farm channel, several furrows can be irrigated at the same time.

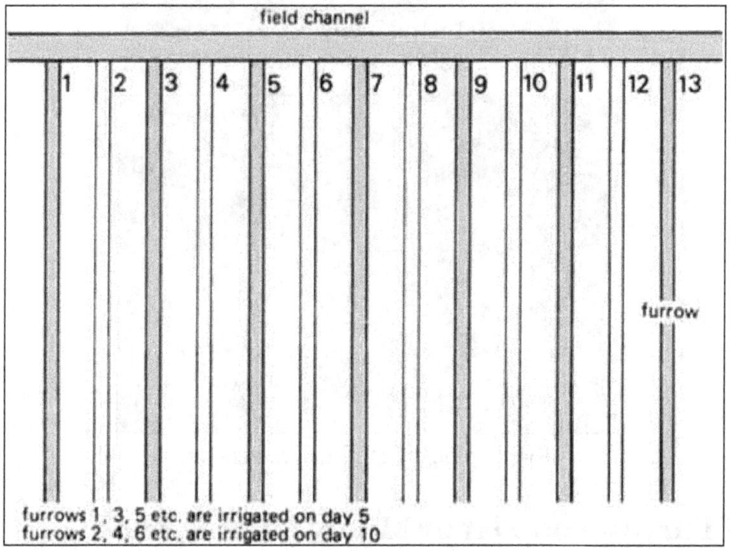

When there is a water shortage, it is possible to limit the amount of irrigation water applied by using 'alternate furrow irrigation'. This involves irrigating alternate furrows rather than every furrow. Figure is an example of this procedure. Instead of irrigating every furrow after 10 days, furrows 1, 3, 5, etc. are irrigated after 5 days and furrows 2, 4 and 6, etc. are irrigated after 10 days. Thus the crop receives some water every 5 days instead of a large amount every 10 days. Small amounts applied frequently in this way are usually better for the crop than large amounts applied after longer intervals of time.

Runoff at the ends of furrows can be a problem on sloping land. This can be as much as 30 percent of the inflow, even under good conditions. Therefore a shallow drain should always be made at the end of the field, to remove excess water. When no drain is made, plants may be damaged by water-logging. Light vegetation allowed to grown in the drain can prevent erosion. Excessive runoff can be prevented by reducing the inflow once the irrigation water has reached the end of the furrows. This is called cut-back irrigation. It may also be possible to reuse runoff water further down the farm.

Wetting Patterns

In order to obtain a uniformly wetted rootzone, furrows should be properly spaced, have a uniform slope and the irrigation water should be applied rapidly.

As the root zone in the ridge must be wetted from the furrows, the downward movement of water in the soil is less important than the lateral (or sideways) water movement. Both lateral and downward movement of water depends on soil type as can be seen in figure.

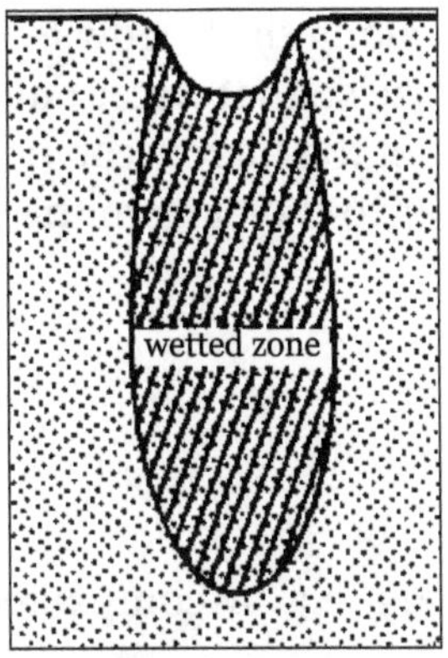

Different wetting patterns in furrows, depending on the soil type (A - SAND).

Different wetting patterns in furrows, depending on the soil type (B - LOAM).

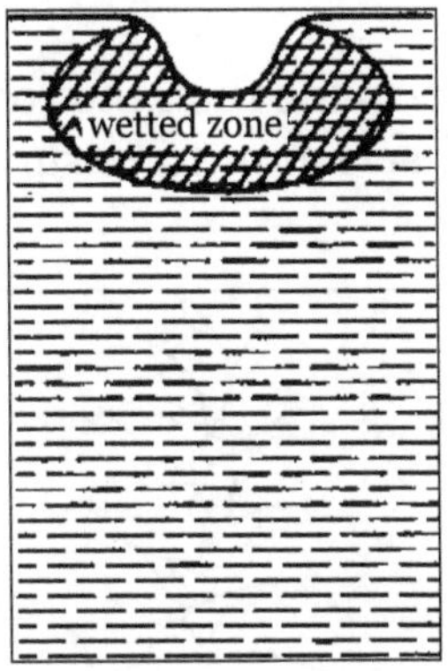

Different wetting patterns in furrows, depending on the soil type (C - CLAY).

Ideal Wetting Pattern

In an ideal situation adjacent wetting patterns overlap each other, and there is an upward movement of water (capillary rise) that wets the entire ridge, thus supplying the root zone with water.

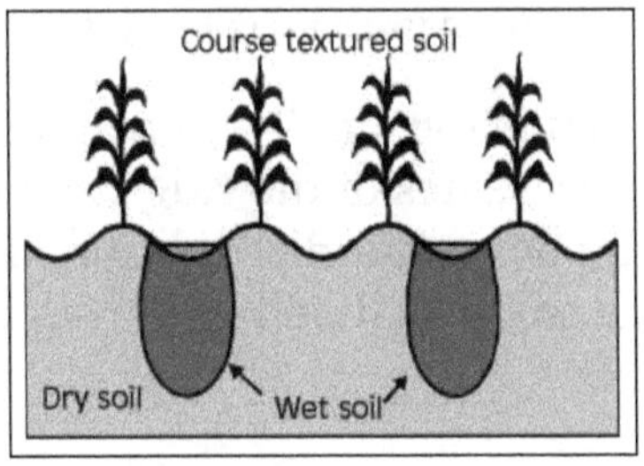

To obtain a uniform water distribution along the furrow length, it is very important to have a uniform slope and a large enough stream size so that water advances rapidly down the furrow. In this way large percolation losses at the head of the furrow can be avoided. The quarter time rule is used to determine the time required for water to travel from the farm channel to the end of the furrow, in order to minimize precolation losses.

Poor Wetting Patterns

Poor wetting patterns can be caused by:

- Unfavourable natural conditions, e.g. a compacted layer, different soil types, uneven slope.
- Poor layout, e.g. a furrow spacing too wide.
- Poor management: supplying a stream size that is too large or too small, stopping the Inflow too soon.

Unfavourable natural conditions: Compacted soil layers or different soil types have the same effect on furrow irrigation as they have on basin irrigation. The solution to the problem is also similar.

An uneven slope can result in uneven wetting along the furrow. Water flows fast down the steep slopes and slowly down the flatter slopes. This affects the time available for infiltration and results in poor water distribution. The problem can be overcome by regrading the land to a uniform slope.

Poor layout: If the furrow spacing is too wide then the root zone will not be adequately wetted. The spacing of furrows needs careful selection to ensure adequate wetting of the entire root zone.

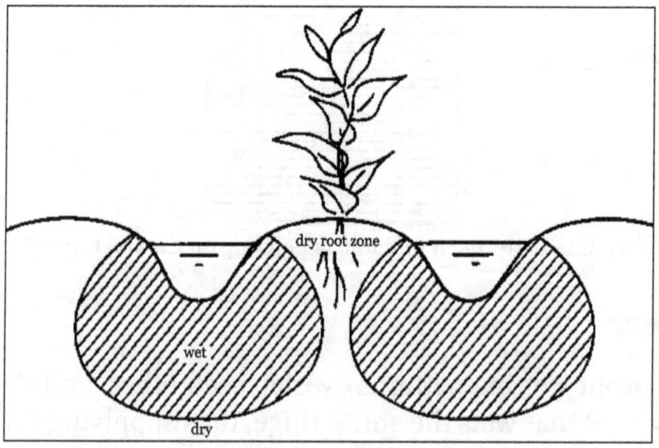

The spacing between two adjacent furrows is too wide.

Poor management: A stream size that is too small will result in inadequate wetting of the ridges. Even if the plants are located at the sides of the ridge, not enough water will be available. A small stream size will also result in poor water distribution along the length of the furrow. The advance will be slow and too much water will be lost through deep percolation at the head of the furrow.

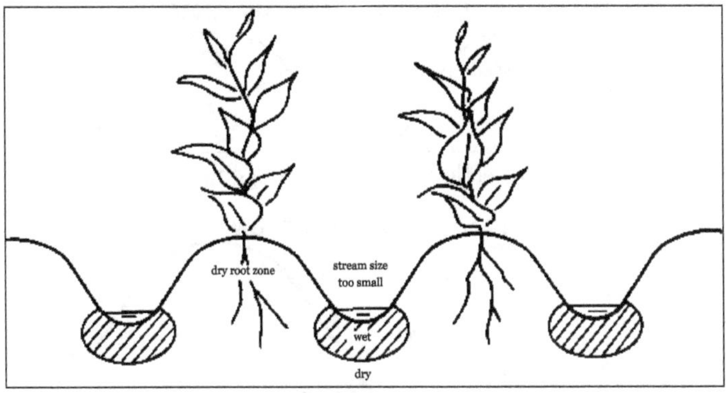

Stream size is too small to wet the ridge.

If the stream size is too large on flat slopes, overtopping of the ridge may occur. On steeper slopes with too large a stream size, erosion of the bed and sides of the furrow may take place.

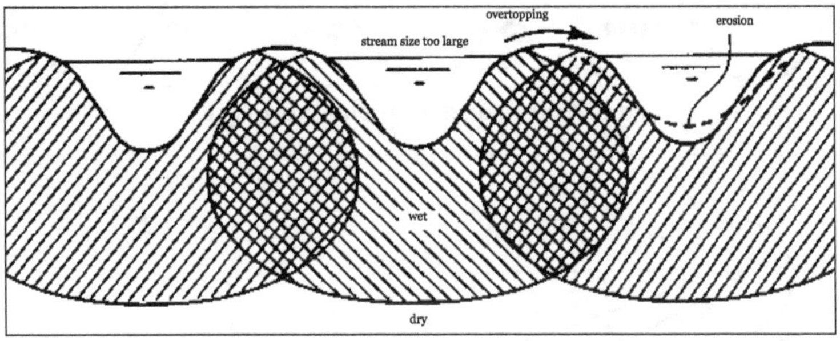

Stream size too large causing overtopping or erosion.

A common management fault is to stop the inflow too soon. This is usually done to reduce runoff, but it results in a poor water distribution and the plants in particular at the end of the furrow do not get enough water. If the inflow of irrigation water is not stopped soon enough, the runoff is excessive and plants at the end of the furrow may drown when an adequate drainage system to evacuate excess water is not provided.

Planting Techniques

The location of plants In a furrow system is not fixed but depends on the natural circumstances. A few examples will be mentioned:

- In areas with heavy rainfall, the plants should stand on top of the ridge in order to prevent damage as a result of waterlogging.
- If water is scarce, the plants may he put in the furrow itself, to benefit more from the limited water.
- As salts tend to accumulate in the highest point, a crop on saline soils should be planted away from the top of the ridge. Usually it is planted in two rows at the sides. However, it is important to make sure there is no danger of waterlogging.
- For winter and early spring crops in colder areas, the seeds may be planted on the sunny side of the ridge. In hotter areas, seeds may be planted on the shady side of the ridge, to protect them from the sun.

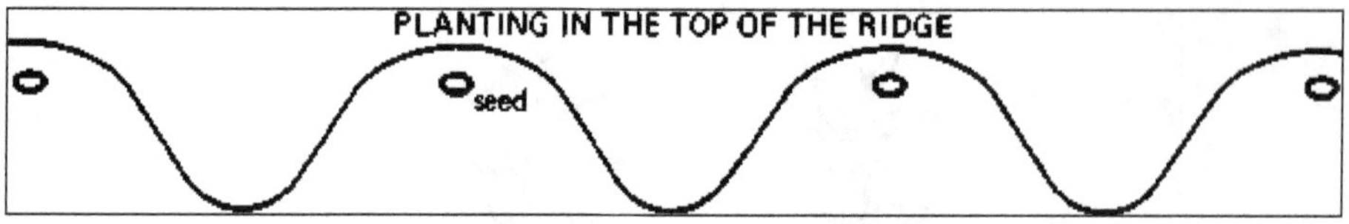

Protection against waterlogging.

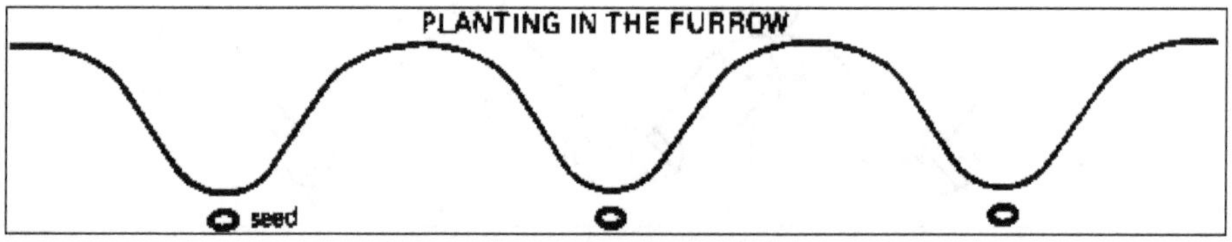

Protection against water scarcity.

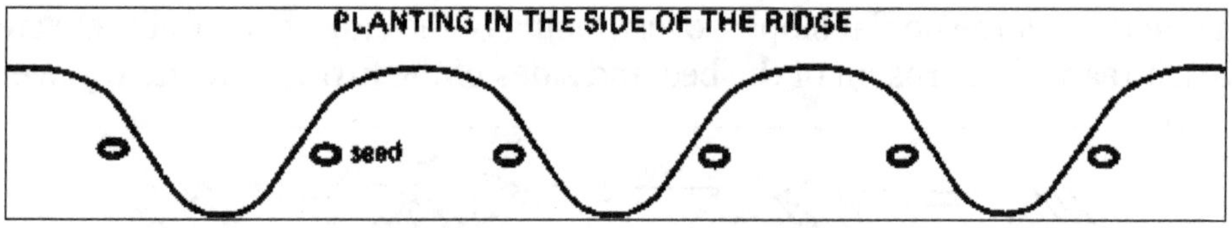

Protection against accumulation of salt.

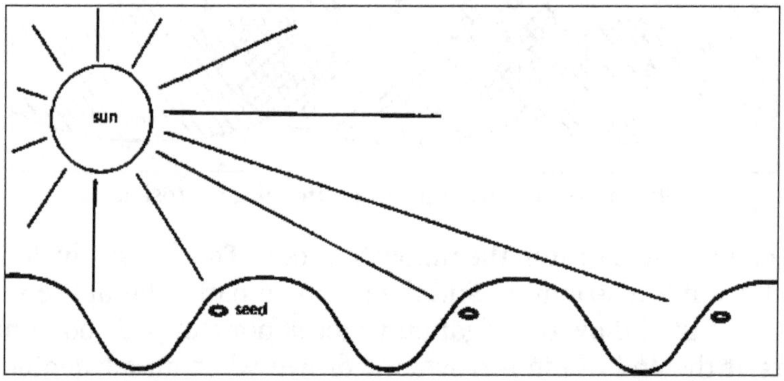

Winter and early spring crops: Seeds planted on the sunny side of the ridge.

Maintenance of Furrows

After construction the furrow system should be maintained regularly; during irrigation it should be checked if water reaches the downstream end of all furrows. There should be no dry spots or places where water stays ponding. Overtopping of ridges should not occur. The field channels and drains should be kept free from weeds.

BORDER IRRIGATION

Border irrigation is generally best suited to the larger mechanized farms as it is designed to produce long uninterrupted field lengths for ease of machine operations. Borders can be up to 800 m or more in length and 3-30 m wide depending on a variety of factors. It is less suited to small-scale farms involving hand labour or animal-powered cultivation methods.

- Suitable slopes: Border slopes should be uniform, with a minimum slope of 0.05% to provide adequate drainage and a maximum slope of 2% to limit problems of soil erosion.

- Suitable soils: Deep homogenous loam or clay soils with medium infiltration rates are preferred. Heavy, clay soils can be difficult to irrigate with border irrigation because of the time needed to infiltrate sufficient water into the soil. Basin irrigation is preferable in such circumstances.
- Suitable crops: Close growing crops such as pasture or alfalfa are preferred.

Border Layout

The dimensions and shape of borders are influenced in much the same way as basins and furrows by the soil type, stream size, slope, irrigation depth and other factors such as farming practices and field or farm size.

Many of the comments made about basins and furrows are generally applicable to borders also and so do not require repetition here. Table provides a guideline to determine maximum border dimensions. It must, however, be stressed that this table is for general guidance only as the values are based on field experience and not on any scientific relationships.

Table: Suggested Maximum Border Lengths and Widths.

Soil type	Border Slope (%)	Unit flow per metre width (l/sec)	Border Width (m)	Border Length (m)
Sand	0.2-0.4	10-15	12-30	60-90
Infiltration rate greater than 25 mm/h	0.4-0.6	8-10	9-12	60-90
	0.6-1.0	5-8	6-9	75
Loam	0.2-0.4	5-7	12-30	90-250
Infiltration rate of 10 to 25 mm/h	0.4-0.6	4-6	6-12	90-180
	0.6-1.0	2-4	6	90
Clay	0.2-0.4	3-4	12-30	180-300
Infiltration rate less than 10 mm/h	0.4-0.6	2-3	6-12	90-180
	0.6-1.0	1-2	6	90

The flow is given per metre width of the border. Thus the total flow into a border is equal to the unit flow multiplied by border width (in metres).

Irrigating Borders

Borders are irrigated by diverting a stream of water from the channel to the upper end of the border. The water flows down the slope. When the desired amount of water has been delivered to the border, the stream is turned off. This may occur before the water has reached the end of the border. There are no specific rules controlling this decision. However, if the flow is stopped too soon there may not be enough water in the border to complete the irrigation at the far end. If it is left running for too long, then water may run off the end of the border and be lost in the drainage system.

As a guideline, the inflow to the border can be stopped as follows:

- On clay soils, the inflow is stopped when the irrigation water covers 60% of the border. If, for example, the border is 100 m long a stick is placed 60 m from the farm channel. When the water front reaches the stick, the inflow is stopped.
- On loamy soils it is stopped when 70 to 80% of the border is covered with water.
- On sandy soils the irrigation water must cover the entire border before the flow is stopped.

However, these are only guidelines. Realistic rules can only be established locally when testing the system.

Wetting Patterns

As with the other irrigation methods it is important to ensure that adequate irrigation water is supplied to the borders so that it fills the root zone uniformly. However, there are many common problems which result in poor water distribution. These include:

- Poor land grading.
- Wrong stream size.
- Stopping the inflow at the wrong time.

1. Poor land grading: If the land is not graded properly and there is a cross-slope, the irrigation water will not spread evenly over the field. It will flow down the slope always seeking the lowest side of the border. This can be corrected by regrading the border to eliminate the cross-slope or by constructing guide bunds in the border to prevent the cross flow of water.

2. Wrong stream size: A stream size which is too small will result in deep percolation losses near the field channel, especially on sandy soils.

If the stream size is too large the water will flow too quickly down the border and the point where the flow should be stopped is reached before sufficient water has been applied to fill the root zone. In this situation the flow will need to be left running until the root zone has been adequately filled and this results in considerable losses from surface runoff. Large stream sizes may also cause soil erosion.

3. Inflow stopped at the wrong time: If the inflow is stopped too soon, the water may not even reach the end of the border. In contrast, if the flow is left running too long, water will run off the border at the downstream end and be lost in the drainage system.

Maintenance of Borders

Maintenance of borders consists of keeping the border free from weeds and uniformly sloping. Whatever damage occurs to the bunds must be repaired and the field channel and drains are to be weeded regularly. By checking frequently and carrying out immediate repairs where necessary, further damage is prevented.

BASIN IRRIGATION

Basin irrigation is suitable for many field crops. Paddy rice grows best when its roots are submerged in water and so basin irrigation is the best method to use for this crop.

Other crops which are suited to basin irrigation include:

- Pastures, e.g. alfalfa, clover.
- Trees, e.g. citrus, banana.
- Crops which are broadcast, such as cereals.
- To some extent row crops such as tobacco.

Basin irrigation is generally not suited to crops which cannot stand in wet or waterlogged conditions for periods longer than 24 hours. These are usually root and tuber crops such as potatoes, cassava, beet and carrots which require loose, well-drained soils.

Suitable Land Slopes

The flatter the land surface, the easier it is to construct basins. On flat land only minor levelling may be required to obtain level basins.

It is also possible to construct basins on sloping land, even when the slope is quite steep. Level basins can be constructed like the steps of a staircase and these are called terraces.

Suitable Soils

Which soils are suitable for basin irrigation depends on the crop grown. A distinction has to be made between rice and non-rice or other crops.

Paddy rice is best grown on clayey soils which are almost impermeable as percolation losses are low. Rice could also be grown on sandy soils but percolation losses will be high unless a high water table can be maintained. Such conditions sometimes occur in valley bottoms.

Although most other crops can be grown on clays, loamy soils are preferred for basin irrigation so that waterlogging (permanent saturation of the soil) can be avoided. Coarse sands are not recommended for basin irrigation as, due to the high infiltration rate, percolation losses can be high. Also soils which form a hard crust when dry (capping) are not suitable.

Basin Layout

Basin layout not only refers to the shape and size of basins but also to the shape and size of the bunds. What is the shape of the basin: square, rectangular or irregular? What is the size of the basin: 10, 100, 1000 or 10 000 m^2? How high should the bund be: 10, 50 or 100 cm? What is the shape of the bund?

Shape and Size of Basins

The shape and size of basins are mainly determined by the land slope, the soil type, the available stream size (the water flow to the basin), the required depth of the irrigation application and farming practices.

Basin Width

The main limitation on the width of a basin is the land slope. If the land slope is steep, the basin should be narrow, otherwise too much earth movement will be needed to obtain level basins. Table provides some guidance on the maximum width of basins or terraces, depending on the land slope.

Three other factors which may affect basin width are:

- Depth of fertile soil,
- Method of basin construction,
- Agricultural practices.

If the topsoil is shallow, there is a danger of exposing the infertile subsoil when the terraces are excavated. This can be avoided by reducing the width of basins and thus limiting the depth of excavation.

Table: Approximate Values for the Maximum Basin or Terrace Width (M).

	Maximum width (m)	
Slope %	average	range
0.2	45	35-55
0.3	37	30-45
0.4	32	25-40
0.5	28	20-35
0.6	25	20-30
0.8	22	15-30
1.0	20	15-25
1.2	17	10-20
1.5	13	10-20
2.0	10	5-15
3.0	7	5-10
4.0	5	3-8

Basins can be quite narrow if they are constructed by hand labour but will need to be wider if machines are used so that the machines can easily be moved around.

If hand or animal powered tillage is used then basins can be much narrower than if machines are used for cultivation. If machines are used then it is important to make sure that basin widths are some multiple of the width of the machines for efficient mechanization.

Basin Size

The size of basins depends not only on the slope but also on the soil type and the available water flow to the basins. The relationship between soil type, stream size and size of the basin is given in table. Values are based on practical experience, and have been adjusted in particular to suit small-scale irrigation conditions.

Table: Suggested Maximum Basin Areas (m²) for Various Soil Types and Available Stream Sizes (l/Sec).

Stream size (l/sec)	Sand	Sandy loam	Clay loam	Clay
5	35	100	200	350
10	65	200	400	650
15	100	300	600	1000
30	200	600	1200	2000
60	400	1200	2400	4000
90	600	1800	3600	6000

Example of How to Estimate Basin Sizes:

Question:	Estimate the dimensions of basins, when the soil type is a deep clay loam and the land slope is 1%. As basin construction is mechanized, the terraces should be as wide as possible. The available stream size is 25 l/sec.
Answer:	The maximum basin or terrace width for a slope of 1% is 25 m (range 15-25 m).
	The maximum basin size for a clay loam soil and an available stream size of 25 l/sec is 1000 m².
	If the total basin area is 1000 m² and the width is 25 m, the maximum basin length is 1000/25 = 40 m.
Note:	This example shows how to estimate the maximum basin dimensions. This basin can be made smaller than this if required and still be irrigated efficiently with the available stream size.

The size of the basin is also influenced by the depth (in mm) of the irrigation application. If the required irrigation depth is large, the basin can be large. Similarly, if the required irrigation depth is small, then the basin should be small to obtain good water distribution.

The size and shape of basins can often be limited by farming practice. Many farms in developing countries are very small and cultivation is by hand. In these circumstances basins are usually small as they are easy to level and efficient irrigation can be attained with relatively small stream sizes.

On the large mechanized farms, basins are generally made as large as possible to provide large uninterrupted areas for machine movements. Basin dimensions are chosen to be some multiple of the width of the machines so as to use the equipment as efficiently as possible. Other reasons to make basins as large as possible are that less land is wasted in this way (less bunds) and large stream sizes and a relatively large application depth can be used.

The shape of the basin can be square, rectangular or irregular. The shape is mainly determined by the slope. On steep and irregular sloping lands, the basins may be long and narrow. The long side of the basin is along the contour line. If the slope and thus the contour line is irregular, the shape of the basin will also be irregular.

Shape and Dimensions of Bunds

Bunds are small earth embankments which contain irrigation water within basins. They are sometimes called ridges, dykes or levees. The height of bunds is determined by the irrigation depth and the freeboard. The freeboard is the height above the irrigation depth to be sure that water will not overtop the bund.

The width of bunds should be such that leakage will not occur, and that they are stable.

Temporary bunds are normally 60-120 cm wide at the base and have a height of 1.5-30 cm above the original ground surface, including a freeboard of 10 cm (which means an irrigation depth of 5-20 cm). Temporary bunds surround fields on which annual crops are grown; these bunds are rebuilt each season.

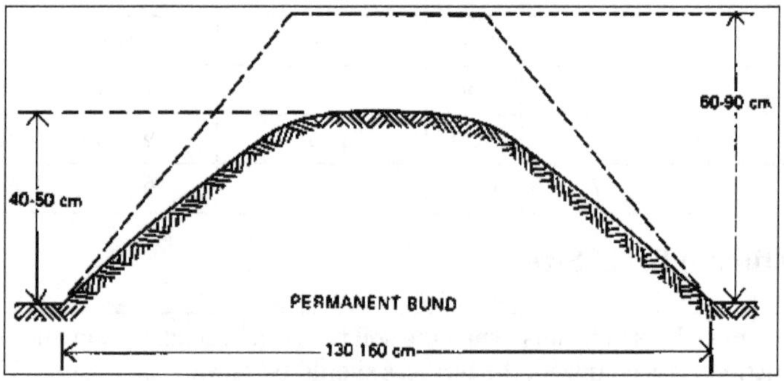

Shape and dimensions of permanent bunds.

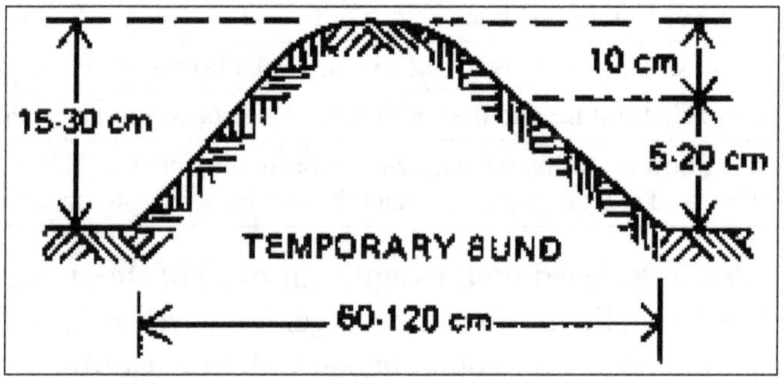

Shape and dimensions of temporary bunds.

Permanent bunds usually have a base width of 130-160 cm and a height of 60-90 cm when constructed. The settled height will be 40-50 cm. This settling (compaction of the soil) will take several months.

Permanent bunds are mostly used in rice cultivation, where the same crop is planted on the same fields year after year. The bunds are used as paths in the rice fields as well. Temporary bunds may be used to subdivide the various fields further, for example as indicated in figure.

Basin Construction

The following steps are involved in the construction of basins: setting out; forming the bunds; and smoothing the land within the basins.

Step 1: Setting Out:

Before construction can begin the location of the basins and bunds must be set out on the ground. This can be done using pegs, string lines or chalk powder to mark the lines of the bunds.

On flat land basins may be square or rectangular in shape. Setting out is relatively simple and involves only straight lines. On sloping or undulating land basins may be irregular in shape and terracing required. Terraces are set out so that the bunds are located along contour lines;

the differences in elevation within each basin should not be excessive so that the amount of earth movement required to obtain a level land surface is small.

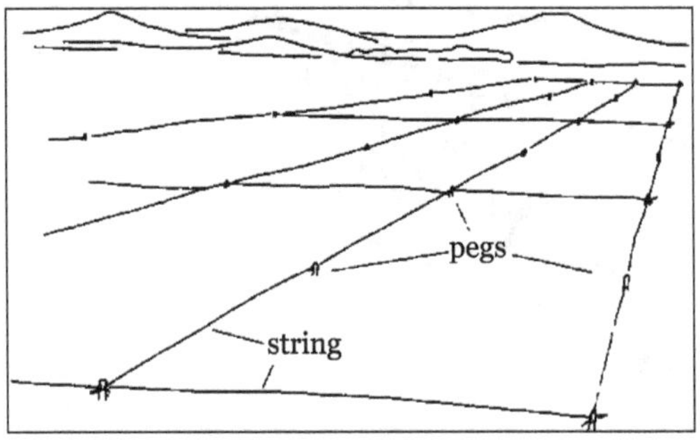

Setting out the markers.

A terrace is set out by first locating a suitable contour line across the land slope. This is the line along which the first bund is constructed. A second line is then set out along a contour further up the slope to mark the location of the next bund.

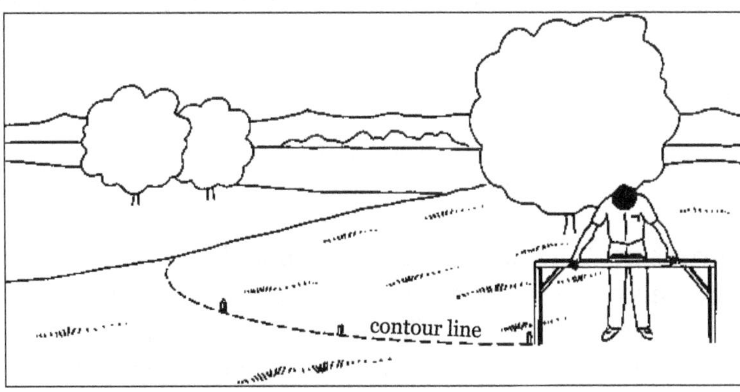

Marking a contour line.

Step 2: Forming the bunds:

Both temporary and permanent bunds can be formed by hand labour or by animal or tractor powered equipment. When soil is gathered from an area close to the bund a 'borrow-furrow' is formed. This furrow can be smoothed out later or be used as a farm channel or drain. When forming bunds for terraces, soil should only be taken from the uphill side of the bund.

A useful piece of equipment for forming bunds is an A-frame. This consists of two boards set on edge and cross-braced, with a wide opening at the front and a narrow opening at the rear. The boards act as blades for cutting into the soil and crowding it into a ridge or bund. A typical A-frame suitable for drawing by animals has blades 20 cm deep and 2 m long spaced 1.5 m apart at the front and 30 cm apart at the rear.

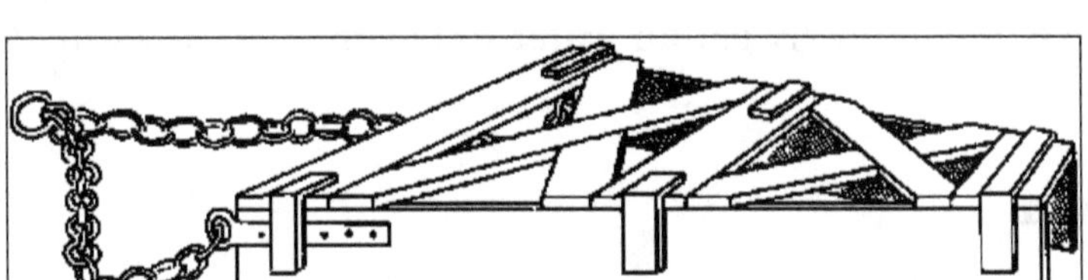

Wooden A-frame.

Making the bunds.

Before forming bunds with an A-frame it is useful to loosen the top soil to a depth of 10-15 cm so that the blades can easily collect sufficient soil.

Whichever method is used it is important that the bunds are properly compacted so that leakage cannot occur.

Step 3: Smoothing the land:

This can be the most difficult part of basin construction and involves very careful levelling of the land within each basin.

On flat land this involves smoothing out the minor high and low spots so that the differences in level are less than 3 cm. This can be done by hand or by a tractor-drawn land plane depending on the size of the basin. However, 3 cm level differences are almost impossible to judge by eye and only when applying water will it become obvious where high and low spots still exist. Thus several attempts may be required to correct the levelling.

Levelling rice basins can be much simpler. These are first cultivated and then filled with water. As the water surface is level, it will be obvious where the high spots are. These can be smoothed put and the water in the basin gradually lowered to reveal other high areas. The smoothing is usually done by an animal or tractor drawn float. This method of smoothing usually destroys the soil structure. This is not a problem when growing rice, but it is not a recommended procedure for other crops.

On sloping land, where terraces are constructed, levelling is achieved by moving soil from the upper part of the slope to the lower part. Care is needed when filling in the borrow furrow to ensure the bund height is maintained so that overtopping is avoided.

Irrigating Basins

There are two methods to supply irrigation water to basins: the direct method and the cascade method.

The Direct Method

Irrigation water is led directly from the field channel into the basin through siphons, spiles or bundbreaks.

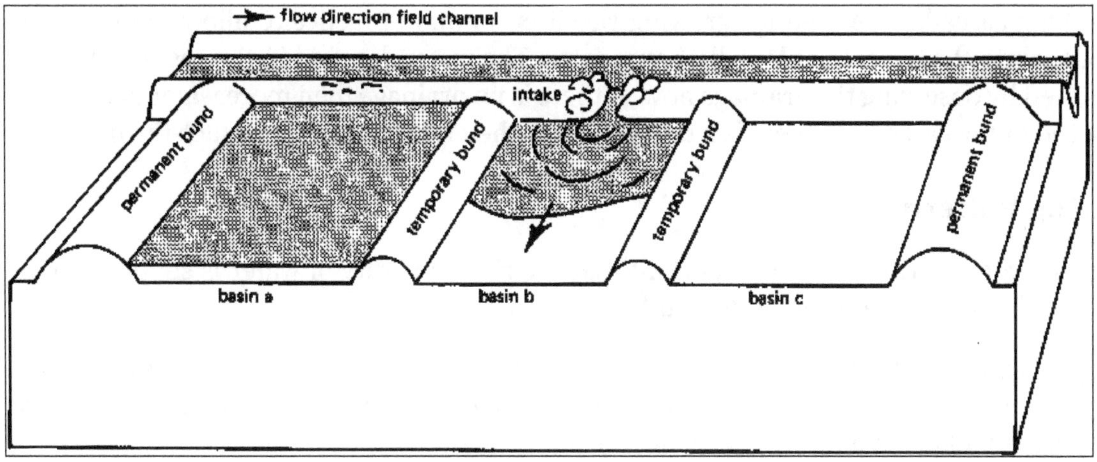

"Basin a" is irrigated first, then "Basin b" is irrigated and so on. This method can be used for most crop types and is suitable for most soils.

The Cascade Method

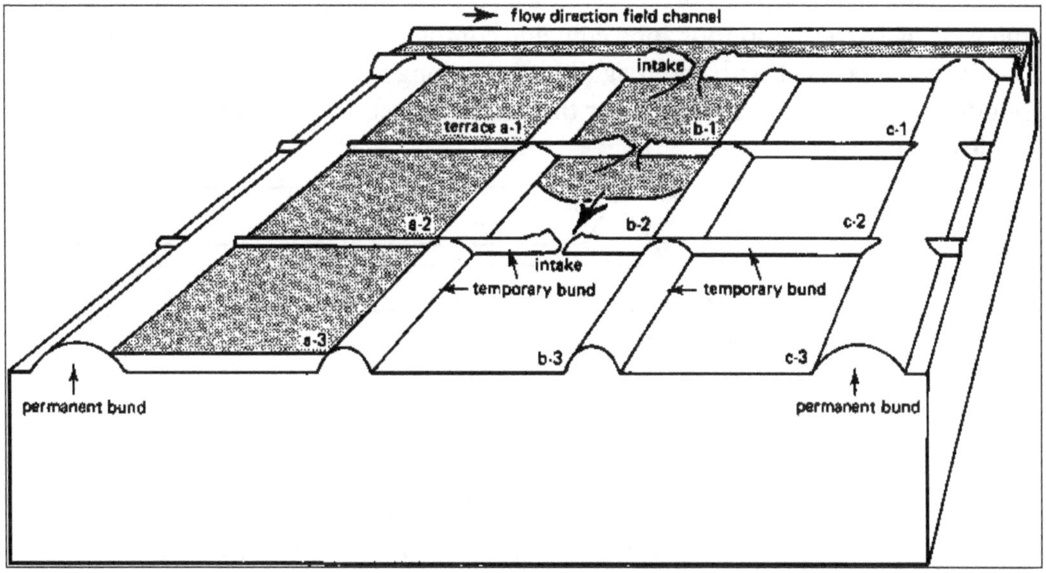

On sloping land, where terraces are used, the irrigation water is supplied to the highest terrace, and then allowed to flow to a lower terrace and so on. In figure the water is supplied to the highest terrace (a.1) and is allowed to flow through terrace a.2 until the lowest terrace (a.3) is filled. The intake of terrace a.1 is then closed and the irrigation water is diverted to terrace b.1 until b.1, b.2 and b.3 are filled, and so on.

This is a good method to use for paddy rice on clay soils where percolation and seepage losses are low. However, for other crops on sandy or loamy soils, percolation losses can be excessive while water is flowing through the upper terraces to irrigate the lower ones. This problem can be overcome by using the borrow-furrow as a small channel to take water to the lower terrace. The lower terrace is irrigated first and when complete the bund is closed and water is diverted into the next terrace. Thus the terrace nearest the supply channel is the last to be irrigated.

When long cascades are used for growing rice it is common practice to allow water to flow continuously into the terraces at low discharge rates. The water demand in the cascade can easily be monitored by observing the drainage flow. If there is no drainage then more water may be required at the top of the cascade. If there is a drainage flow then it is possible to reduce the inflow.

Wetting Patterns

For good crop growth it is very important that the right quantity of water is supplied to the root zone and that the root zone is wetted uniformly.

If crops receive too little water, they will suffer from drought stress, and yield may be reduced. If they receive too much water, then water is lost through deep percolation and, especially on clay soils, permanent pools may form, causing the plants to drown.

Ideal Wetting Pattern

To obtain a uniformly wetted root zone, the surface of the basin must be level and the irrigation water must be applied quickly. Ideal wetting pattern: the basin is level and the right quantity of water has been supplied with the correct scream size. As can be seen from figure it is not possible to have the wetting pattern and root zone coincide completely. The part of the basin near the field channel is always in contact with the irrigation water longer than the opposite side of the basin. Therefore percolation losses will occur near the field channel, if sufficient water is supplied to the opposite side of the basin.

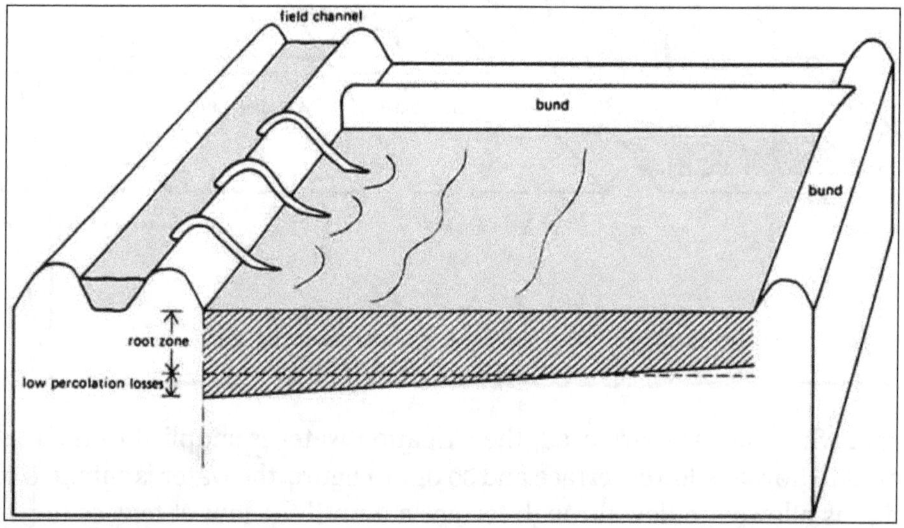

Poor Wetting Patterns

Poor wetting patterns can be caused by:

- Unfavourable natural conditions, e.g. a compacted subsoil layer, or different soil types within one basin;
- Poor layout, e.g. a poorly levelled surface;
- Poor management, e.g. supplying incorrect stream size, applying too little or too much water.

Unfavourable natural conditions: A compacted sub-soil layer can sometimes occur in a basin some 30-50 cm below the soil surface. Infiltration through this layer may be very slow and so water tends to accumulate above this layer: a "perched" water table is formed. This may result in waterlogging.

This situation may be very helpful for growing rice but will be harmful for other crops. The compacted layer can be removed by using deep ploughs or rippers which break up the subsoil.

Different soil types within a basin can cause very uneven water distribution. This problem can be solved by re-aligning basin boundaries so that each basin contains only one soil type.

Poor layout: Some parts of the root zone receive too little water and in the lower parts water may pond or be lost through deep percolation. Plants suffer in the drier parts because they receive too little water and wilt. Plants may also suffer in the wet parts; plant nutrients are carried away from the rootzone to the subsoil and, especially on clay soils, the plants may drown. These faults can easily be corrected by careful land levelling.

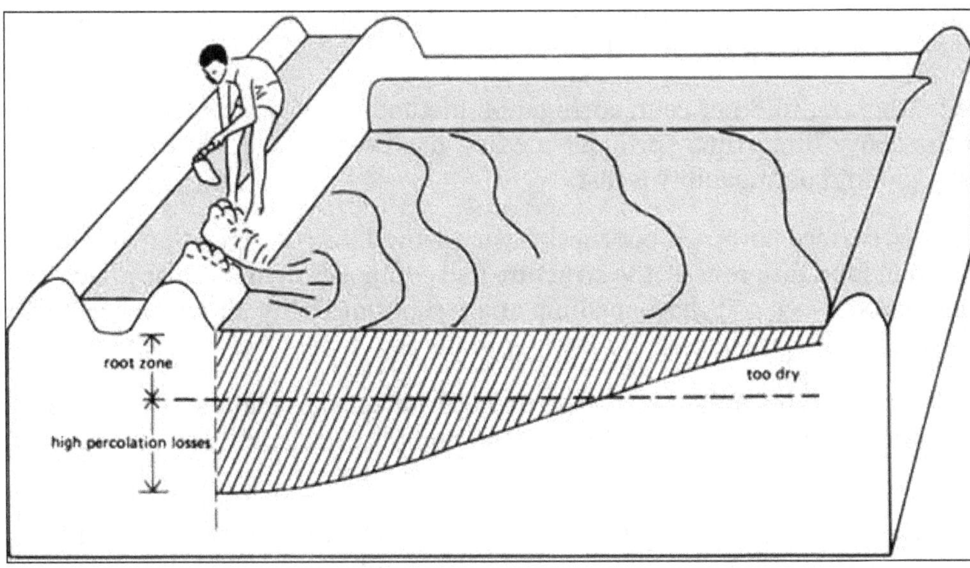

Poor management: If the basin is irrigated too slowly, by using a stream size which is too small. The part of the basin which receives irrigation water first (near the supply channel) and thus the longest, receives too much water. Percolation losses occur, nutrients are washed away and the plants may drown. The other end of the basin remains too dry. The plants there do not receive enough water and wilt.

The solution to the problem is to:

- Increase the stream size so that the basin will be flooded more rapidly,
- Subdivide the basin into smaller basins; smaller basins need a smaller stream size than larger basins.

There are no percolation losses during under-irrigation. Although water may be used efficiently by this approach, frequent irrigation will be necessary to meet crop water needs. However, continual under-irrigation will eventually restrict root development and the crop may suffer when there are delays in irrigating, e.g. when water is in short supply or the supply system breaks down.

Maintenance of Basins

Bunds are susceptible to erosion which may be caused by, for example, rainfall, flooding or the passing of people when used as footpaths. Rats may dig holes in the sides of the bunds. It is therefore important to check the bunds regularly, notice defects and repair them instantly, before greater damage is done. Before each growing season, the basins should be checked to see that they remain level. During pre-irrigation it can easily be seen where higher and lower spots are; there should be smoothed out. Also, the field channels should be kept free from weeds and silt deposits.

FACTORS AFFECTING METHODS OF IRRIGATION

The following factors affect any selected method of irrigation:

1. Topography: Topography is the slope of the ground and how much uneven or leveled it is. The irrigation method is selected accordingly.

If the slope is from 0.4 to 8 per cent, corrugation method of irrigation is suitable. If the slope is more than the above limit, then sprinkler method has been considered more suitable for soils which are shallow and permeability is fast.

Similarly, leveled surface flow method, check basin method, furrow irrigation method etc., are adopted. Apart from level differences, the structure and composition of soil also affects the selection of irrigation method. Deep soils have medium and fast permeability for which corrugated method of irrigation is suitable. Rate of seepage of water in the soil depends on its composition.

2. Climate: Ground level irrigation methods are directly affected by climate. The sprinkler system of irrigation is most suitable for such climate. Dryness, humidity and speed of air affect the methods of irrigation.

3. Means of Irrigation: The sources of irrigation and the chemical composition of the water also affect the irrigation method. At the time of irrigation by tube wells, the sprinkler method or drip irrigation method is suitable, but in canal irrigation, flood method is selected.

4. Crops: The irrigation method is selected according to type of crops and pattern of its sowing. Different crops require different quantities of water. Growth of plants and their height are also affected. In taller plants, check basin irrigation method is more suitable as compared to sprinkler method.

5. Texture of the soils crusting, cracking and infiltration characteristics of the surface soil; nature and depth of relatively impermeable layer(s) in the sub-soil, if any; the water storage capacity of the potential root zone; the magnitude and nature of land slope; the length and size of the field; surface drainage; the nature and extent of salts in the surface soil and the sub-soil, also decide the method of irrigation.

6. Water Conservation: Water conservation is the demand of the day when the whole world is facing water crisis. Hence, such a method of irrigation should be adopted which uses minimum water but provides maximum humidity to the plants. Sprinkler method and drip irrigation method are the best from the view point of water conservation.

7. Economic Factors and Labour: While selecting any method of irrigation, the economic condition must be kept in mind as many irrigation methods require heavy initial investment but lesser investment later on. Surface irrigation methods require continuous investment. Along with investment, availability of labour is also an essential factor. As compared to sprinkler method, check basin method of irrigation requires more labour.

8. General: Outlook, management skills and financial resources of the farmer; nature of farm machinery used; availability and cost of labour; maintenance facilities and costs of irrigation equipment; availability of power supply should also be considered.

As far as possible, an irrigation method should not only provide a high level of water application efficiency but also ensure its economic viability, sustained soil productivity and wide adaptability to the prevalent features of the farm.

References

- Traditional-methods-of-irrigation, biology: byjus.com, Retrieved 4 June, 2019
- Mader, Shelli (May 25, 2010). "Center pivot irrigation revolutionizes agriculture". The Fence Post Magazine. Retrieved June 6, 2012
- Traditional-and-modern-irrigation-methods: biologydiscussion.com, 5 July, 2019
- Manual-irrigatio, optimisation-water-use-agriculture, hardware, water-use, implementation-tools, category: archive.sswm.info, Retrieved 7 August, 2019
- Bainbridge, David A (June 2001). "Buried clay pot irrigation: a little known but very efficient traditional method of irrigation". Agricultural Water Management. 48 (2): 79–88. Doi:10.1016/S0378-3774(00)00119-0
- Traditional-and-modern-irrigation-methods: biologydiscussion.com, Retrieved 9 January, 2019
- "US Patent # 5,387,307. Drip irrigation tape and method of manufacture - Patents.com". Www.patents.com. Retrieved 2017-09-30
- Spring-irrigatio, agricultural-engineering: tnau.ac.in, Retrieved 5 February, 2019
- R. Goyal, Megh (2012). Management of drip/trickle or micro irrigation. Oakville, CA: Apple Academic Press. P. 104. ISBN 978-1926895123
- Localized-irrigation: akvopedia.org, Retrieved 19 March, 2019
- Drainage Manual: A Guide to Integrating Plant, Soil, and Water Relationships for Drainage of Irrigated Lands. Interior Dept., Bureau of Reclamation. 1993. ISBN 0-16-061623-9
- The-following-factors-affect-any-selected-method-of-irrigation: biologydiscussion.com, Retrieved 20 April, 2019

3

Water Sources for Irrigation

Various water sources that are used for irrigation are ground water, tube well, surface water, irrigation canals and tanks. Some of the non-conventional sources are also used for irrigation including desalinated water, spate irrigation and rain harvesting. This chapter closely examines these water sources of irrigation to provide an extensive understanding of the subject.

GROUNDWATER

Groundwater is the water that occurs below the surface of Earth, where it occupies all or part of the void spaces in soils or geologic strata. It is also called subsurface water to distinguish it from surface water, which is found in large bodies like the oceans or lakes or which flows overland in streams. Both surface and subsurface water are related through the hydrologic cycle (the continuous circulation of water in the Earth-atmosphere system).

Most groundwater comes from precipitation. Precipitation infiltrates below the ground surface into the soil zone. When the soil zone becomes saturated, water percolates downward. A zone of saturation occurs where all the interstices are filled with water. There is also a zone of aeration where the interstices are occupied partially by water and partially by air. Groundwater continues to descend until, at some depth, it merges into a zone of dense rock. Water is contained in the pores of such rocks, but the pores are not connected and water will not migrate. The process of precipitation replenishing the groundwater supply is known as recharge. In general, recharge occurs only during the rainy season in tropical climates or during winter in temperate climates. Typically, 10 to 20 percent of the precipitation that falls to the Earth enters water-bearing strata, which are known as aquifers.

Groundwater is constantly in motion. Compared to surface water, it moves very slowly, the actual rate dependent on the transmissivity and storage capacity of the aquifer. Natural outflows of groundwater take place through springs and riverbeds when the groundwater pressure is higher than atmospheric pressure in the vicinity of the ground surface. Internal circulation is not easily determined, but near the water table the average cycling time of water may be a year or less, while in deep aquifers it may be as long as thousands of years.

Groundwater plays a vital role in the development of arid and semiarid zones, sometimes supporting vast agricultural and industrial enterprises that could not otherwise exist. It is particularly fortunate that aquifers antedating the formation of deserts remain unaffected by increases in aridity with the passage of time. Withdrawal, however, will deplete even the largest of groundwater basins so that development based on the existence of aquifers can be only temporary at best.

A vast amount of groundwater is distributed throughout the world, and a large number of groundwater reservoirs are still underdeveloped or uninvestigated. Scientists estimate that some 5.97 quintillion gallons (22.6 million cubic km [5.4 million cubic miles]) of groundwater reside in the upper 2 km (1.2 miles) of Earth's surface. The most frequently investigated or exploited groundwater reservoirs are of the unconsolidated clastic (mainly sand and gravel) or carbonate hardrock type found in alluvial valleys and coastal plains under temperate or arid conditions.

Though some groundwater dissolves substances from rocks and may contain traces of old seawater, most groundwater is free of pathogenic organisms, and purification for domestic or industrial use is not necessary. Furthermore, groundwater supplies are not seriously affected by short droughts and are available in many areas that do not have dependable surface water supplies. However, aquifers and other groundwater supplies are at risk of chemical pollution from fracking, agricultural chemicals, leaking or unfit landfills and septic tanks, and other point and nonpoint sources of pollution. Such contamination can render groundwater unfit for use and is expensive and difficult to clean up.

TUBE WELL

A tube well used for irrigating wheat.

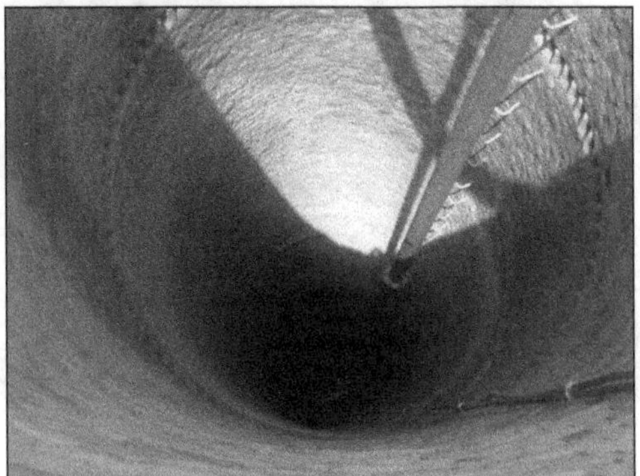

Borehole of a tube well, which is usually 30 metres (98 ft) deep or more.

A tube well is a type of water well in which a long, 100–200 millimetres (3.9–7.9 in)-wide, stainless steel tube or pipe is bored into an underground aquifer. The lower end is fitted with a strainer, and a pump lifts water for irrigation. The required depth of the well depends on the depth of the water table.

Temporary Reservoir

A small reservoir of water is made at the outlet of the tube well. This reservoir is used for different usage of water by the local population.

Casing

The tube well casing houses the inlet, cylinder, piston valves and rising main of a "down-the-hole" type hand pump. Casing to support the external surfaces of the borehole against collapse may be needed, either temporarily or permanently, and is often made of PVC pipe, which is both cheap and inert.

Seepage down the tube well bore is prevented by the sanitary seal. Seepage from the ground above the aquifer is excluded by the lengths of plain casing. Water to be pumped is admitted through slots in the lower lengths of casing.

Water abstracted from aquifers in relatively soft ground usually contains sand or silt particles, which are liable to cause rapid wear to pump valves and cylinders (and dissatisfaction among consumers). Methods of preventing these particles from reaching the pump are of two general types, screening and sand/gravel packing.

Screening

The drawing shows slots in the PVC casing which can be cut on site, using a hacksaw and it is sealed Glass Reinforces Epoxy (GRE). Strainers have also been widely used in projects in Bangladesh.

Sand/Gravel Packing

Outlet of a tube well to a temporary reservoir.

The drawing shows graded sand and gravel, which is placed from the top of the borehole. More compact, pre-bonded, packs of sand and gravel are available commercially; some of these can also form part of the pump inlet assembly. Sand and gravel packing is meant to eliminate particles from the water before they reach the screen and that would otherwise have passed through.

Ground Water Safety

The introduction of tube wells has led to major arsenic poisoning in Bangladesh. High concentrations of arsenic occur naturally in the soil in some areas of Bangladesh.

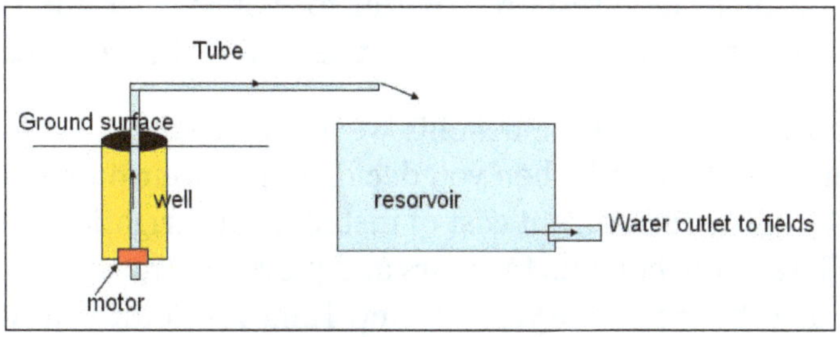

Diagram of a tube well.

Advantages of Tube Well Irrigation

- Groundwater is easily available.
- The water table is fairly close to the surface.
- Able to irrigate a much larger area.
- More reliable during periods of drought when surface water dries up.
- Suitable for small holdings.

IRRIGATION WELL

An irrigation well uses water from your well to power a variety of systems that distribute non-potable water, such as sprinkler systems, animal waterers, geo systems, and feeding ponds, among others. Some companies use water from a shallower depth, since it won't be used for human consumption, as a way to cut corners and avoid regulations.

What are the Benefits of Irrigation Wells?

Irrigation wells have many benefits for all areas of your life, including your home, your wallet, and your environment:

- Lower utility bills: Because well water is free, you don't have to worry about a sky-high water bill in the summer, when your lawn, garden, and outdoor plants need extra water to stay green and healthy. While home owners will often see some savings, for commercial enterprises, such as farms, an irrigation well is practically a necessity. Reliable access to affordable water can make or break an industry, and an irrigation well is one way to ensure that your agricultural business succeeds.
- Avoid unexpected price hikes: Because you're using your own water from your own well, you don't have to worry about blowing your budget on unexpected price surges. Free is free.

- Healthy, clean water: Well water doesn't contain chlorine or other chemicals. This is especially good news for those of us with gardens or farms, who prefer to eat organically, or who want to at least know exactly what is in the water they'll eventually consume.
- Groundwater is a renewable resource: Better for your wallet and better for the environment? It sounds too good to be true, but trust us—this is the real deal.
- Increase in property value: All the benefits we just listed are great for you, but they're also wonderful selling points if and when you decide to put your home on the market. This is a great way to rationalize the upfront cost of installing an irrigation well. By spending a little extra now, you'll reap the benefits for years and years to come. In Pender County, you'll get your return on investment in one year. In New Hanover County, it will take 1.25 years, and in Onslow County, 1.5 years.
- A time saver: Time is our most precious commodity, and if you have a large lawn, an impressive garden, or a farm, then watering it by hand can be anything from annoying to impossible. An irrigation well simplifies the process and can even be set to run by a timer.

SURFACE WATER

Image of the entire surface water flow of the Alapaha River near Jennings, Florida going into a sinkhole leading to the Floridan Aquifer groundwater.

Surface water is water on the surface of continents such as in a river, lake, or wetland. It can be contrasted with groundwater and atmospheric water.

Non-saline surface water uses is replenished by precipitation and by recruitment from ground-water. It is lost through evaporation, seepage into the ground where it becomes ground-water, used by plants for transpiration, extracted by mankind for agriculture, living, industry etc. or discharged to the sea where it becomes saline.

Conjunctive use of Ground and Surface Water

Surface and groundwater are two separate entities, so they must be regarded as such. However, there is an ever-increasing need for management of the two as they are part of an interrelated system that is paramount when the demand for water exceeds the available supply. Depletion of surface and ground water sources for public consumption (including industrial, commercial, and residential) is caused by over-pumping. Aquifers near river systems that are over-pumped have been known to deplete surface water sources as well. Research supporting this has been found in numerous water budgets for a multitude of cities.

Response times for an aquifer are long. However, a total ban on ground water usage during water recessions would allow surface water to retain better levels required for sustainable aquatic life. By reducing ground water pumping, the surface water supplies will be able to maintain their levels, as they recharge from direct precipitation, surface runoff, etc.

MAIN IRRIGATION CANAL

Main irrigation canal is the canal that delivers water to the lands to be irrigated and is composed of the off-load part (from the head works to the head of distribution canals) and working part where distributing conducting canals branch off from the main canal.

Main canal is the major part of the conductive network, because water is delivered through it from the source to the entire irrigated territory. The main canal is composed of the head section, and off-load and working sections.

Head section is the inlet water-intake section where water is cleared from sediments if sediment tanks are available. The canal's off-load part delivers water from an irrigation source to the first distribution canal. The canal's working section distributes water between distribution canals.

The main canal is laid orienting on the highest points of a given irrigated area.

An irrigated land can be supplied with water from a source by gravity or by pump irrigation. At gravity water supply, water intake can be of dam intake and river-canal (damless) intake types. At that, water comes from a river (reservoir) through a check structure to the off-load part of the main canal by which it is transported to a particular irrigated land. At pumping irrigation, water comes through pipes.

The location and length of the main canal's off-load section depend on the slope of the ground and main canal, water withdrawal source water level rise above the earth's surface where the water needs to be taken to.

In some cases, minimum slope does not provide cost-effective solution, since the cross section becomes too large and considerably increases the volume of earthwork operations. and so the most advantageous slope is established, at which no silting will happen and the main canal construction cost will be lowest.

When laying canal, the command point to which water is required to be delivered from a river by gravity or a water withdrawal point. In the first case, it would be necessary to determine the length of the main canal's off-load section and a water withdrawal point. Canal laying is carried out from the command point towards the river. In the second case, it would be necessary to determine the length of the off-load section of the main canal and determine the upper boundary of the command area. The canal is laid from a water withdrawal point towards the irrigated area under design.

The main canal must command the entire irrigated area, from the end of the off-load section to the end of the last cluster of structures from which only a distribution network is constructed.

Depending on the relief, one can outline three key layouts of the main canal and inter-farm network. The main canal is routed towards the horizontals (across the slope) at a design gradient along the upper boundary of the respective irrigated area. First-order distributors are designed as double-sided command type along the maximum slope. If there are hollows, uplands and subdivides, the main canal is designed based on the same principle; however, along the upper boundary it loses its straightness. Distribution canals are routed at an allowable gradient along the dividing crests and to ensure commanding both slopes.

The main canal is laid along the maximum surface slope, approximately in the middle of the designed irrigated land. The distributors are laid at minimum allowable gradient along the land contours. Such a layout of canals is reasonable to apply for the irrigated lands with long length along the slope (gradient) and narrow width. In this case, drop structures are designed only on the main canal, while in the first layout version those have to be designed on all first-order distributors. At considerable flow rate in the main canal, a hydropower plant is constructed in the places of concentrated water fall.

Under the conditions of complex relief, the main canal, inter-farm, and farm distributors are arranged on command points and watershed divides so that the flow velocities through canals should be lower than the scouring (erosive) velocity and higher than the silting velocity and the requirements to the management of such an irrigated area as well as other conditions of canal designing should be met.

IRRIGATION TANK

An irrigation tank or tank is an artificial reservoir of any size. They are mainly found in India. (The word *sagar* refers to a large lake, usually man-made). It can also have a natural or man-made spring included as part of a structure. Tanks are part of an ancient tradition of harvesting and preserving the local rainfall and water from streams and rivers for later use, primarily for agriculture and drinking water, but also for sacred bathing and ritual. Often a tank was constructed across a slope so to collect and store water by taking advantage of local mounds and depressions. Tank use is especially critical in parts of South India without perennial rainfall where water supply replenishment is dependent on a cycle of dry seasons alternating with monsoon seasons.

Tank Design

Water is considered a purifying and regenerative element in India, and is an essential element of prayer and ritual. Water is also revered because of its scarcity in western India where dry and monsoon seasons alternate and failure of the monsoon season means famine and death while plentiful water replacing irrigations sources is a time of rejoicing. This resulted in building water storage tanks that combined the practical and sacred. Since ancient times, the design of water storage has been important in Indian architecture. As early as 3000 BC, sophisticated systems of drains, wells and tanks were built to conserve and utilise water. Tank building as an art form began with the Hindus and developed under Muslim rule.

An example of the art of tank design is the large, geometically spectacular Stepped Tank at the Royal Center at the ruins of Vijayanagara, the capital of the Vijayanagara Empire, surrounding the modern town of Hampi. It is lined with green diorite and has no drain. The tank was filled by aqueduct.

Village Tanks

A tank.

Ralegaon Siddhi is an example of a village that revitalised its ancient tank system. In 1975 the village was drought-stricken. The village tank could not hold water as the earthen embankment dam wall leaked. Work began with the percolation tank construction by the villagers who donated their labor to repair the embankment. Once this was fixed, the village's seven wells below the tank filled with water in the summer for the first time in memory. Now the village has a supply of water throughout the year.

Temple Tanks

Tanks known as "Pushkarni" or "Kalyani" also known as "Kund" in Hindi are reservoirs with steps leading down to the water, generally found in South India, and sometimes constructed within the walls of a temple complex.

Bathing in the sacred waters of a temple tank was believed to cure worshippers of afflictions such as leprosy and blindness. Many temple tanks are decaying and drying up today.

Stepwells

Stepwells, also called *bavdi* are wells in which the water can be reached by descending a set of steps. They are most common in the west of India.

Stepwells were often used for leisure, providing relief from daytime heat. This led to the building of some significant ornamental and architectural features, often associated with dwellings and in urban areas. It also ensured their survival as monuments.

Modern Tank Management

The development of large-scale water management methods and hydroelectric generation have replaced much of the local efforts and community management of water. For example, the state of Karnataka has about 44,000 artificial wetlands locally constructed over many centuries. At least 328 are threatened today.

However, recently a tank regeneration movement initiated by communities and non-governmental organisations (NGO) has arisen. Today, there are approximately 120,000 small-scale tanks, irrigating about 41,200 km^2 in semi-arid areas of India. This constitutes about one third of the total irrigated land in South India.

NON-CONVENTIONAL SOURCES

Desalinated Water

Desalinated water is being used for irrigation purposes in many places around the world, where the available water source is too saline to be used. Such water sources may include groundwater, sea water and sometimes surface water from lakes.

Water salinity is a major problem for agriculture. Saline water contains high concentrations of salts, which crops might not tolerate.

The source of the salts in water is the surrounding rocks, e.g. the mineralogy of the aquifer. Therefore, surface water is usually less salty than groundwater, which is surrounded by rocks. Oceans are salty because when water evaporates, the salts remain and accumulate.

Saline water affects crops in two ways:

1. It increases the osmotic pressure and reduces the ability of the plant to absorb water, even up to a point in which the plant wilts.
2. High concentrations of specific ions in the saline water may be toxic to the plant.

Groundwater may contain high concentrations of ions such as sodium, chlorides, sulphate, bicarbonates, calcium, magnesium. It may also contain trace elements such as boron, iron, manganese and fluoride, which may be present at relatively low concentrations, but might become toxic to the plant if their concentrations exceed certain thresholds (such thresholds are usually crop-specific).

Desalination is a water treatment process in which salts are removed from water. There are various methods of desalination, of which, the main method used today is Reverse Osmosis.

In reverse osmosis, water is pushed through semi-permeable membranes, using pressure. The salts do not pass through the membrane, while water molecules do.

Although there is some selectivity to specific ions, desalination with reverse osmosis removes most salts from the water and the amount of salts removed depends mostly on the pressure and on the ratio between the amount of desalinated water produced and the amount of brine water rejected and discharged as waste.

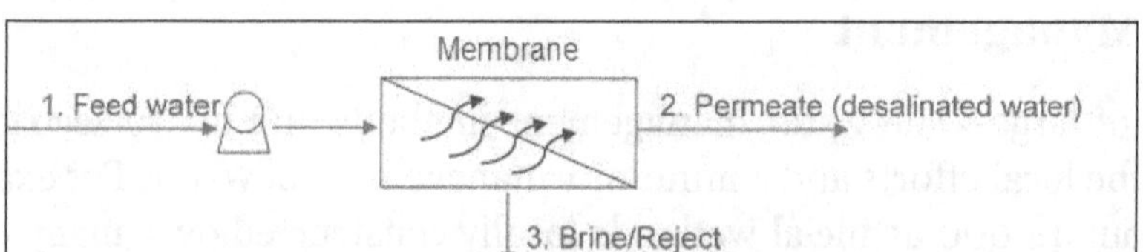

Flow rate of feed water = flow rate of permeate + flow rate of brine.

Higher rejected brine volume results in better quality of the desalinated water and vice versa.

As we can see, desalination removes most ions (salts) from the treated water. Many water sources (especially groundwater) may contain elements that are considered essential plant nutrients. Such elements include calcium, magnesium and sulfur, which are required in relatively large quantities by plants.

Removing these salts, together with the harmful elements, such as sodium and chlorides, requires adding back those elements with fertilizers. Such fertilizers may include calcium nitrate, magnesium nitrate and magnesium sulphate.

Desalination also removes carbonate hardness from water. This results in water that has a lower resistance to changes in pH. Adding even a small amount of acid to the desalinated water might result in a sharp drop of its pH. Therefore, in such cases, the water has to be stabilized by adding sources of bicarbonate, such as potassium bicarbonate.

Under some circumstances the level of harmful salts can be compromised, depending on the concentration of such salts in the raw water and the tolerance level of the crop. In such cases, one of the two following options can be applied, depending on the specific conditions and water quality:

- The reverse osmosis plant can be operated at higher permeate flow rates, leaving more salts in the desalinated water, i.e. leaving higher concentration of harmful salts, but also higher level of essential nutrients.
- Desalinated water can be mixed with untreated water.

Applying the above methods, can save energy and fertilizer costs.

It is important to run a simulation of the reverse osmosis plant, prior to its final design and installation. The manufacturer of the plant will usually run such a simulation, but it is the grower's responsibility to provide a full laboratory analysis of the raw water and to define the required water quality and flow rates. Except of salts, does reverse osmosis remove other substances?

Since dissolved salts, which are removed by reverse osmosis, are much smaller than viruses, reverse osmosis also removes viruses and bacteria.

Spate Irrigation

Spate irrigation uses seasonal floods of rivers, streams, ponds and lakes to fill water storage canals. If irrigation is the manipulation of water for use in growing crops, spate irrigation is perhaps the most ancient method devised. Evidence of basic spate irrigation engineering can be traced back beyond recorded history and include the Ancient Egyptian diversion of the Nile River for storage at times of heavy rainfall for use in the dryer times of the year.

In spate irrigation, water is diverted from normally dry river beds when the river is in spate. The flood water is then diverted to the fields. This may be done by free intakes, by diversion spurs or by bunds, that are built across the river bed. The flood water, typically lasting a few hours or a few days, is channeled through a network of primary, secondary and sometimes tertiary flood channels. Command areas may range from anything between a few hectares to over 25,000 hectares (62,000 acres).

Spate water management is more common to arid and semi-arid regions, typically in underdeveloped countries in the Middle East, North Africa, West Asia, East Africa and parts of Latin America, but is used in countries with sufficient precipitation to manage water elevation in canals.

Spate irrigation systems require huge management efforts to control and optimize the flow of water. Because fast-moving water is capable of moving very large amounts of sediment, the heights of river banks and the composition of their beds can change rapidly. Diverting the flow of a powerful force which is capable of moving rocks, soils and other materials used to divert the path of the water can prove difficult.

The number of public programs to support spate irrigation has been relatively limited. One reason has been the difficulty to justify investments in civil engineering works on systems dominated with low-value farming. A second reason has been that it has been hard to identify successful interventions in spate systems, because spate systems are often hydraulically and socially generally very complex.

Rain Harvesting

Rainwater capture and storage system at the Monterrey Institute of Technology and Higher Education, Mexico City.

Rainwater harvesting is the accumulation and storage of rainwater for reuse on-site, rather than allowing it to run off. Rainwater can be collected from rivers or roofs, and in many places, the water collected is redirected to a deep pit (well, shaft, or borehole), a reservoir with percolation, or collected from dew or fog with nets or other tools. Its uses include water for gardens, livestock, irrigation, domestic use with proper treatment, indoor heating for houses, etc. The harvested water can also be used as drinking water, longer-term storage, and for other purposes such as groundwater recharge.

Rainwater harvesting is one of the simplest and oldest methods of self-supply of water for households usually financed by the user.

Current uses

Canada

A number of Canadians have started implementing rainwater harvesting systems for use in stormwater reduction, irrigation, laundry, and lavatory plumbing. Substantial reform to Canadian law since the mid-2000s has increased the use of this technology in agricultural, industrial, and residential use, but ambiguity remains amongst legislation in many provinces. Bylaws and local municipal codes often regulate rainwater harvesting.

India

- Tamil Nadu was the first state to make rainwater harvesting compulsory for every building to avoid groundwater depletion. The scheme was launched in 2001 and has been implemented in all rural areas of Tamil Nadu. Posters all over Tamil Nadu including rural areas create awareness about harvesting rainwater TN Govt site. It gave excellent results within five years, and slowly every state took it as a role model. Since its implementation, Chennai had a 50% rise in water level in five years and the water quality significantly improved.

- Karnataka: In Bangalore, adoption of rainwater harvesting is mandatory for every owner or the occupier of a building having the site area measuring 60 ft (18.3 m) × 40 ft (12.2 m) and above and for newly constructed building measuring 30 ft (9.1 m) × 40 ft (12.2 m) and above dimensions. In this regard, Bangalore Water Supply and Sewerage Board has initiated and constructed "Rain Water Harvesting Theme Park" in the name of Sir M. Visvesvaraya in 1.2 acres (4,900 m^2) of land situated at Jayanagar, Bangalore. In this park, 26 different type of rainwater harvesting models are demonstrated along with the water conservation tips. The auditorium on the first floor is set up with a "green" air conditioning system and will be used to arrange the meeting and showing of a video clip about the rainwater harvesting to students and general public. An attempt has been made at the Department of Chemical Engineering, IISc, Bangalore to harvest rainwater using upper surface of a solar still, which was used for water distillation.

- In Rajasthan, rainwater harvesting has traditionally been practised by the people of the Thar Desert. Many ancient water harvesting systems in Rajasthan have now been revived. Water harvesting systems are widely used in other areas of Rajasthan, as well, for example the *chauka* system from the Jaipur district.

- Maharashtra: At present, in Pune, rainwater harvesting is compulsory for any new housing society to be registered.

- In Mumbai, Maharashtra, rainwater harvesting is not mandatory. The Mumbai City council is planning to make rainwater harvesting mandatory for large societies.

Israel

The Southwest Center for the Study of Hospital and Healthcare Systems in cooperation with Rotary International is sponsoring a rainwater harvesting model program across the country. The first rainwater catchment system was installed at an elementary school in Lod, Israel. The project is looking to expand to Haifa in its third phase. The Southwest Center has also partnered with the Water Resources Action Project of Washington, DC, which currently has rainwater harvesting projects in the West Bank. Rainwater harvesting systems are being installed in local schools for the purpose of educating schoolchildren about water conservation principles and bridging divides between people of different religious and ethnic backgrounds, all while addressing the water scarcity issue that the Middle East faces.

New Zealand

Although New Zealand has plentiful rainfall in the West and South, for much of the country, rainwater harvesting is the normal practice for most rural housing and is encouraged by most councils.

Sri Lanka

Rainwater harvesting has been a popular method of obtaining water for agriculture and for drinking purposes in rural homes. The legislation to promote rainwater harvesting was enacted through the Urban Development Authority (Amendment) Act, No. 36 of 2007. Lanka rainwater harvesting forum is leading the Sri Lanka's initiative.

South Africa

The South African Water Research Commission has supported research into rainwater harvesting. Reports on this research are available on their "Knowledge Hub". Studies in arid, semiarid, and humid regions have confirmed that techniques such as mulching, pitting, ridging, and modified run-on plots are effective for small-scale crop production. hydrofracturing has regularly been used to improve the performance of water boreholes. From 1990 to 1992, 170 boreholes had been hydrofractured.

United Kingdom

In the United Kingdom, water butts are often found in domestic gardens and on allotments to collect rainwater, which is then used to water the garden. However, the British government's Code For Sustainable Homes encouraged fitting large underground tanks to newly built homes to collect rainwater for flushing toilets, watering, and washing. Ideal designs had the potential to reduce demand on mains water supply by half. The code was revoked in 2015.

Other Countries

- In China, Argentina, and Brazil, rooftop rainwater harvesting is being mayank for livestock, water for small irrigation, and a way to replenish groundwater levels. Gansu province in China and semiarid northeast Brazil have the largest rooftop rainwater harvesting projects going on.
- Thailand has the largest fraction of the population in the rural area relying on rainwater harvesting (currently around 40%). Rainwater harvesting was promoted heavily by the government in the 1980s. In the 1990s, after government funding for the collection tanks ran out, the private sector stepped in and provided several million tanks to private households, many of which continue to be used. This is one of the largest examples of self-supply of water worldwide.

Rainwater harvesting tank.

- In Bermuda, the law requires all new construction to include rainwater harvesting adequate for the residents.
- The U.S. Virgin Islands has a similar law.
- In Senegal and Guinea-Bissau, the houses of the Diola-people are frequently equipped with homebrew rainwater harvesters made from local, organic materials.
- In the Irrawaddy Delta of Myanmar, the groundwater is saline and communities rely on mud-lined rainwater ponds to meet their drinking water needs throughout the dry season. Some of these ponds are centuries old and are treated with great reverence and respect.
- In the United States, until 2009 in Colorado, water rights laws almost completely restricted rainwater harvesting; a property owner who captured rainwater was deemed to be stealing it from those who have rights to take water from the watershed. Now, residential well owners who meet certain criteria may obtain a permit to install a rooftop precipitation collection system (SB 09-080). Up to 10 large scale pilot studies may also be permitted (HB 09-1129). The main factor in persuading the Colorado Legislature to change the law was a 2007 study that found that in an average year, 97% of the precipitation that fell in Douglas County, in the southern suburbs of Denver, never reached a stream—it was used by plants or evaporated on the ground. Rainwater catchment is mandatory for new dwellings in Santa Fe, New Mexico. Texas offers a sales tax exemption on the purchase of rainwater harvesting equipment. Both Texas and Ohio allow the practice even for potable purposes. Oklahoma passed the Water for 2060 Act in 2012, to promote pilot projects for rainwater and graywater use among other water-saving techniques.
- In Beijing, some housing societies are now adding rainwater in their main water sources after proper treatment.
- In Ireland, Professor Micheal Mcginley established a project to design a rainwater harvesting prototype in the biosystems design challenge module at University College Dublin.

New Approaches

Presentation of Rain Saucer system to students at orphanage in Guatemala.

Instead of using the roof for catchment, the Rain Saucer, which looks like an upside-down umbrella, collects rain straight from the sky. This decreases the potential for contamination and makes potable water for developing countries a potential application. Other applications of this free-standing rainwater collection approach are sustainable gardening and small-plot farming.

A Dutch invention called the Groasis Waterboxx is also useful for growing trees with harvested and stored dew and rainwater.

Traditionally, stormwater management using detention basins served a single purpose. However, optimized real-time control lets this infrastructure double as a source of rainwater harvesting without compromising the existing detention capacity. This has been used in the EPA headquarters to evacuate stored water prior to storm events, thus reducing wet weather flow while ensuring water availability for later reuse. This has the benefit of increasing water quality released and decreasing the volume of water released during combined sewer overflow events.

Generally, check dams are constructed across the streams to enhance the percolation of surface water into the subsoil strata. The water percolation in the water-impounded area of the check dams can be enhanced artificially manyfold by loosening the subsoil strata and overburden using ANFO explosives as used in open cast mining. Thus, local aquifers can be recharged quickly using the available surface water fully for use in the dry season.

Non-traditional

In 1992, American artist Michael Jones McKean created an artwork in Omaha, Nebraska, at the Bemis Center for Contemporary Art that created a fully sustainable rainbow in the Omaha skyline. The project collected thousands of gallons of rainwater, storing the water in six daisy-chained 12,000 gallons tanks. The massive logistical undertaking, during its five-month span, was one of the largest urban rainwater harvesting sites in the American Midwest.

Rainwater Harvesting by Freshwater-flooded Forests

Rainwater harvesting is possible by growing freshwater-flooded forests without losing the income from the used, submerged land. The main purpose of the rainwater harvesting is to use the locally available rainwater to meet water requirements throughout the year without the need of huge capital expenditure. This would facilitate the availability of uncontaminated water for domestic, industrial, and irrigation needs.

Ratagul Freshwater Flooded Forest, Bangladesh.

Rainwater Harvesting by Solar Power Panels

Good quality water resource, closer to populated areas, is becoming scarce and costly for the consumers. In addition to solar and wind energy, rain water is major renewable resource of any land. Vast area is being covered by solar PV panels every year in all parts of the world. Solar panels can also be used for harvesting most of the rain water falling on them and drinking quality water, free from bacteria and suspended matter, can be generated by simple filtration and disinfection processes as rain water is very low in salinity. Exploitation of rain water for value added products like bottled drinking water, makes solar PV power plants profitable even in high rainfall/cloudy areas by the augmented income from value added drinking water generation.

Advantages

Rainwater harvesting provide the independent water supply during regional water restrictions, and in developed countries, is often used to supplement the main supply. It provides water when a drought occurs, can help mitigate flooding of low-lying areas, and reduces demand on wells which may enable groundwater levels to be sustained. It also helps in the availability of potable water, as rainwater is substantially free of salinity and other salts. Applications of rainwater harvesting in urban water system provides a substantial benefit for both water supply and wastewater subsystems by reducing the need for clean water in water distribution systems, less generated stormwater in sewer systems, and a reduction in stormwater runoff polluting freshwater bodies.

A large body of work has focused on the development of life cycle assessment and its costing methodologies to assess the level of environmental impacts and money that can be saved by implementing rainwater harvesting systems.

Independent Water Supply

Rainwater harvesting provides an independent water supply during water restrictions. In areas where clean water is costly, or difficult to come by, rainwater harvesting is a critical source of clean water. In developed countries, rainwater is often harvested to be used as a supplemental source of water rather than a main source, but the harvesting of rainwater can also decrease a household's water costs or overall usage levels. Rainwater is safe to drink. Rainwater is also independent of salinity or pollutants found in ground water, increasing the quantity of portable drinking water available when rainwater harvesting is utilized.

Supplemental in Drought

When drought occurs, rainwater harvested in past months can be used. If rain is unpredictable, the use of a rainwater harvesting system can be critical to capturing the rain when it does fall. Many countries with arid environments, use rainwater harvesting as a cheap and reliable source of clean water. To enhance irrigation in arid environments, ridges of soil are constructed to trap and prevent rainwater from running down hills. Even in periods of low rainfall, enough water is collected for crops to grow. Water can be collected from roofs and can be constructed to hold large quantities of rainwater.

In addition, rainwater harvesting decreases the demand for water from wells, enabling groundwater levels to be further sustained rather than depleted.

Life-cycle Assessment

Life-cycle assessment is a methodology used to evaluate the environmental impacts of a system from cradle-to-grave of its lifetime. Devkota et al, developed such a methodology for rainwater harvesting, and found that the building design (e.g., dimensions) and function (e.g., educational, residential, etc.) play critical roles in the environmental performance of the system. The Economic and Environmental Analysis of Sanitations Technologies, EEAST model evaluates the greenhouse gas emissions and cost of such systems over the lifetime of a variety of building types.

To address the functional parameters of rainwater harvesting systems, a new metric was developed - the demand to supply ratio (D/S) - identifying the ideal building design (supply) and function (demand) in regard to the environmental performance of rainwater harvesting for toilet flushing. With the idea that supply of rainwater not only saves the potable water, but also saves the stormwater entering the combined sewer network (thereby requiring treatment), the savings in environmental emissions were higher if the buildings are connected to a combined sewer network compared to separate one.

System Setup

Rainwater harvesting systems can range in complexity, from systems that can be installed with minimal skills, to automated systems that require advanced setup and installation. The basic rainwater harvesting system is more of a plumbing job than a technical job, as all the outlets from the building's terrace are connected through a pipe to an underground tank that stores water.

Systems are ideally sized to meet the water demand throughout the dry season, since it must be big enough to support daily water consumption. Specifically, the rainfall capturing area such as a building roof must be large enough to maintain adequate flow of water. The water storage tank size should be large enough to contain the captured water. For low-tech systems, many low-tech methods are used to capture rainwater: rooftop systems, surface water capture and pumping the rainwater that has already soaked into the ground or captured in reservoirs and storing it in tanks (cisterns).

Before a rainwater harvesting system is built, use of digital tools is useful. For instance, to detect if a region has a high rainwater harvesting potential, rainwater-harvesting GIS maps can be made using an online interactive tool, or, to estimate how much water is needed to fulfill a community's water needs, the Rain is Gain tool helps. Tools like these can save time and money before a commitment to build a system is undertaken, in addition to making the project sustainable and long lasting.

Applications of Rainwater Harvesting

Agriculture

Missions to five Caribbean countries have shown that the capture and storage of rainwater runoff for later use is able to significantly reduce the risk of losing some or all of the year's harvest because of soil or water scarcity. In addition, the risks associated with flooding and soil erosion during high rainfall seasons would decrease. Small farmers, especially those farming on hillsides, could benefit the most from rainwater harvesting because they are able to capture runoff and decrease the effects of soil erosion.

Many countries, especially those with arid environments, use rainwater harvesting as a cheap and reliable source of clean water. To enhance irrigation in arid environments, ridges of soil are constructed to trap and prevent rainwater from running down hills and slopes. Even in periods of low rainfall, enough water is collected for crops to grow. Water can be collected from roofs, dams and ponds can be constructed to hold large quantities of rainwater so that even on days when little to no rainfall occurs, enough is available to irrigate crops.

Domestic Use

- In China, Argentina, and Brazil, rooftop rainwater harvesting is used to provide drinking water, domestic water, water for livestock, water for small irrigation, and a way to replenish groundwater levels. Gansu province in China and semiarid northeast Brazil have the largest rooftop rainwater harvesting projects going on.
- About 40% of Thailand's rural population utilizes rainwater harvesting. Rainwater harvesting was promoted heavily by the government in the 1950s. In the 1990s, after government funding for the collection tanks ran out, the private sector stepped in and provided several million tanks to private households, many of which continue to be used today. This is one of the largest examples of self-supply of water worldwide.
- Rainwater harvesting is mandatory for new homes built in Santa Fe, New Mexico.
- Texas offers a sales tax exemption for the purchase of rainwater harvesting equipment.
- Both Texas and Ohio allow rainwater harvesting to be used even for potable purposes.
- Oklahoma passed the Water for 2060 Act in 2012, to promote pilot projects for rainwater and graywater use among other water-saving techniques.
- In the United Kingdom, water butts are often found in domestic gardens and on allotments to collect rainwater, which is then used to water the garden.

Industry

Frankfurt Airport has the biggest rainwater harvesting system in Germany. The system helps save approximately 1 million cubic meters of water per year. The cost of the system was 1.5 million dm (US$63,000) in 1993. This system collects water from roofs of the new terminal which has an area of 26,800 square metres. The water is collected in the basement of the airport in six tanks with a storage capacity of 100 cubic meters. The water is mainly used for toilet flushing, watering plants and cleaning the air conditioning system.

Rainwater harvesting was adopted at The Velodrome – The London Olympic Park – in order to increase the sustainability of the facility. A 73% decrease in potable water demand by the park was estimated. Despite this, it was deemed that rainwater harvesting was a less efficient use of financial resources to increase sustainability than the park's blackwater recycling program.

Quality of Water Harvesting

Rainwater may need to be analyzed properly, and used in a way appropriate to its safety. In the Gansu province for example, solar water disinfection is used by boiling harvested rainwater in parabolic solar cookers before being used for drinking. These so-called "appropriate technology" methods provide low-cost disinfection options for treatment of stored rainwater for drinking.

While rainwater itself is a clean source of water, often better than groundwater or water from rivers or lakes, the process of collection and storage often leaves the water polluted and non-potable. Rainwater harvested from roofs can contain human, animal and bird feces, mosses and lichens, windblown dust, particulates from urban pollution, pesticides, and inorganic ions from the sea (Ca, Mg, Na, K, Cl, SO_4), and dissolved gases (CO_2, NO_x, SO_x). High levels of pesticide have been found in rainwater in Europe with the highest concentrations occurring in the first rain immediately after a dry spell; the concentration of these and other contaminants are reduced significantly by diverting the initial flow of run-off water to waste. Improved water quality can also be obtained by using a floating draw-off mechanism (rather than from the base of the tank) and by using a series of tanks, withdraw from the last in series. Prefiltration is a common practice used in the industry to keep the system healthy and ensure that the water entering the tank is free of large sediments.

Conceptually, a water supply system should match the quality of water with the end use. However, in most of the developed world, high-quality potable water is used for all end uses. This approach wastes money and energy and imposes unnecessary impacts to the environment. Supplying rainwater that has gone through preliminary filtration measures for nonpotable water uses, such as toilet flushing, irrigation and laundry, may be a significant part of a sustainable water management strategy.

EFFECTS OF EXCESSIVE AND IRREGULAR IRRIGATION

As growth and development of plants is prevented due to scarcity of water, similarly excessive water also has an adverse effect on growth of plants.

Due to excess water in soil, voids are filled with water or water collects on the surface of soil. In such situation, balance of air and water is disturbed in pores and in the absence of air, plants are not able to breathe etc.

Hence, neither irrigation should be more nor less but it should be as per requirement because excessive irrigation is dangerous.

Excessive and irregular irrigation has the following effects:

1. Increase in Saline and Alkaline Elements in Soil or Increase in Salinity.

When irrigation is done regularly with excessive water, groundwater level comes up and quantity of solvent salts increases. Due to evaporation, salts from saline groundwater collect on the surface of the soil. Seepage of canal water also results in more salts in irrigation water. As observed in Rajasthan, concentration of soluble salts is so much in these soils that it has an adverse effect on the growth of crops and the production is lesser. When concentration of soluble salts in the soil is more than 0.1, growth of plants stops.

2. Problem of Water logging.

Excessive irrigation also causes problem of water logging as has happened in Indira Gandhi Canal area in Rajasthan. Soils and ground level of Indira Gandhi Canal area is very sensitive from the angle of irrigation as maximum soil there has bold granules and at a little depth below the surface, there are hard rocks of calcium carbonate which do not allow excess water to infiltrate underground but keeps it filled above.

Hence, the problem of waterlogging occurs there. Due to maximum dampness in Naurangdesar, Silvana, Lakhuwali and other areas near Indira Gandhi Main Canal, agriculture activity has almost completely stopped. Pilibanga and Badopal have also similar problem. Thus, 44,625 hectare land in the north of the canal has become waterlogged.

3. Hindrance in Air Communication.

Excessive water replaces air in pores of the soil. Hence, roots of plants and beneficial organisms do not get sufficient air for breathing and living. Hence, growth of plants is stopped and diseases destroy crops.

4. Reduction in Temperature of Soil.

Excess collected water becomes cold in the process of evaporation and it causes reduction in temperature of the soil. Seeds do not settle properly and growth of plants is affected. Activeness of organisms is also reduced.

5. Shallowness of Roots.

Due to excess water and shortage of air, roots do not go deep and they remain on the upper surface. Plants are not able to get nutritive elements in sufficient quantity and shallow roots, because of weakness, are neither able to develop nor are they able to bear swift winds. Due to this the plants fall on the ground.

6. Marshy Land.

Land becomes marshy due to stagnant water for a long duration and only forest grass can grow there. The land does not remain useful for agriculture.

7. More Nitrate Formation.

Due to absence of oxygen in waterlogged soil, process of de-nitrogenization starts. Organisms themselves start drawing available oxygen in nitrate, which convert nitrate into nitrus oxide, nitric oxide and nitrogen gases. Being in gas form, they are useless for plants. Such problems will not exist by balanced irrigation.

8. Shortage of Soil Nutritive Elements and Decrease in Productivity.

Due to excessive irrigation, nutritive elements of soil flow to sub-soil, resulting in decrease in soil productivity and deficiency in crops.

9. Acidity of Soil.

In excessively irrigated areas, H^+ replaces positively charged Ca^2, Mg^{2+}, K^+, Na^+ stick to colloid granules of soil. These positively charged particles flow away to lower levels which cause decrease of basic positively charged particles and there is an excess of colloidal granules. This results in excess of H+ in soil water and soil becomes acidic.

10. Deformation comes in soil structure.

11. Attack of diseases and harmful insects.

12. It encourages the process of leaching.

13. Collection of other harmful salts and alkaline takes place on surface of the soil.

14. There is decrease in activities of living organisms.

References

- Groundwater, science: britannica.com, Retrieved 21 June, 2019
- Palanisami, K, and Easer, E.W. (2000). Tank Irrigation in the 21st Century—What Next? Discovery, Delhi, Discovery. ISBN 81-7141-558-X
- The-benefits-of-irrigation-wells: appliedresourcemgmt.com, Retrieved 25 July, 2019
- Rural Water Supply Network. “Rural Water Supply Network Self-supply site”. Www.rural-water-supply.net/en/self-supply. Retrieved 2017-03-19
- Irrigation-with-desalinated-water: cropaia.com, Retrieved 25 August, 2019
- Devkota, Jay; Schlachter, Hannah; Apul, Defne (May 2015). “Life cycle based evaluation of harvested rainwater use in toilets and for irrigation”. Journal of Cleaner Production. 95: 311–321. Doi:10.1016/j.jclepro.2015.02.021
- Excessive-and-irregular-irrigation-has-the-following-effects: biologydiscussion.com, Retrieved 5 January, 2019

4

Irrigation Management

The practice of monitoring and managing the rate, timing and volume of water according to the crop needs is referred to as irrigation water management. Irrigation scheduling, irrigation system maintenance, etc. are some of the methods used for irrigation management. This chapter has been carefully written to provide an easy understanding of irrigation water management.

IRRIGATION WATER MANAGEMENT

Irrigation Water Management (IWM) is the practice of monitoring and managing the rate, volume, and timing of water application according to the seasonal crop needs, giving consideration to the soil intake and water holding capacities. Soil moisture should be managed to obtain optimum yields, without deep percolation losses or runoff.

Irrigation water management will help irrigators determine the effectiveness of irrigation practices, make good water management decisions, and justify making irrigation adjustment in existing systems.

WATER MANAGEMENT AND PRICING

The most important physical elements of an *irrigation project* are *land* and *water*. In accordance with the propriety relations of these elements there may be different types of water management:

- The communal type.
- The enterprise type.
- The utility type.

Communal Type

Until the end of the 19th century the development of irrigation projects occurred at a mild pace, reaching a total area of some 50 million ha worldwide, which is about 1/5 of the present area. The land was often private property or assigned by the village authorities to male or female farmers, but the water resources were in the hands of clans or communities who managed the water resources *cooperatively*.

Enterprise Type

Slave labor in a cotton plantation.

Labor in a sugarcane plantation.

The enterprise type of water management occurred under large landowners or agricultural corporations, but also in centrally controlled societies. Both the land and water resources are inonehand. Large plantations were found in colonised countries in Asia, Africa, and Latin America, but also in countries employing slave labor. It concerned mostly the large scale cultivation of commercial crops such as bananas, sugarcane and cotton. As a result of land reforms, in many countries the estates were reformed into a cooperatives in which the previous employers became members and exercised a cooperative form of land and water management.

Utility Type

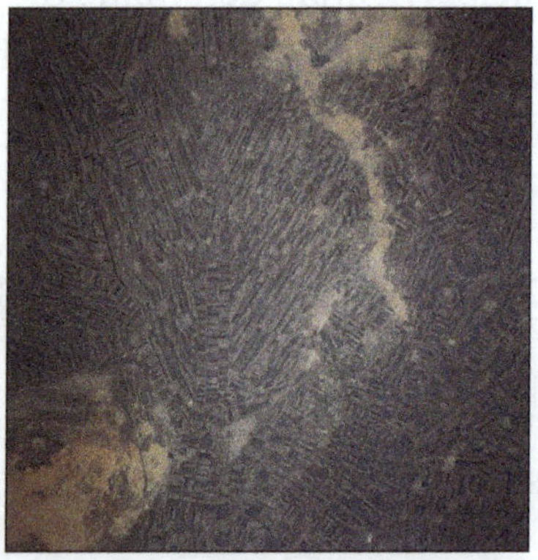

Irrigation canals of the Gezira Scheme, Sudan, from space, 1997, with the utility type of management. The water comes from the Blue Nile.

The utility type of water management occurs in areas where the land is owned by many, but the exploitation and distribution of the water resources are managed by (government) organizations. After 1900 governments assumed more influence over irrigation because:

- Water was increasingly considered government property owing to the increasing demand for good quality water and the reducing availability.
- Governments embarked on large scale irrigation projects as they were considered more efficient.

- The development of new irrigation schemes became technically, financially and organizationally so complicated that they fell outside the capabilities of the smaller communities.
- The import and export policies of governments required the cultivation of commercial cash crops whilst, by controlling the water management, the farmers could be more easily guided to plant these kind of crops.

The water management signified a large subsidy on irrigation schemes. From 1980 the operation and maintenance of many irrigation projects was gradually handed over to water user organizations (WUA's) who were to assume these tasks and a large part of the costs, whereby the water rights of the members had to be respected.

The exploitation of water resources via large storage dams that often provided electric power as well and diversion weirs normally remained the responsibility of the government, mainly because environmental protection and safety issues were at stake.

In the past, the utility type of water management witnessed more conflicts and disturbances then the other types.

Water Pricing

Tariffs

Irrigation water has a price by which the management costs must be covered. The following tariff (water charge) systems exist:

- No tariff, the government assumes the costs.
- Tariff in labor hours, which holds mainly in communal types of management in traditional irrigation systems.
- Yearly area tariff, a fixed price per ha per year.
- Seasonal area tariff, a fixed price per ha per season with the higher price in the dry season.
- Volumetric tariff, a fixed price per m^3 of water; the consumption is measured by water meters.
- Block or stepped-up pricing for water use per ha; the price increases as the water consumption per ha falls in a higher block.

The use of groundwater for irrigation is often licensed by government and the well owner may be permitted to withdraw only a maximum volume of water per year at a certain price.

Cost Recovery

The recovery of water charges may be below target, because:

- The revenues accrue to a (government) organization other than the one responsible for the management.
- Farmers and water users have no say in the water management.
- Lack of communication between farmers and project managers.
- Poor farmers are unable to comply.

- Farmers do not receive water according to need; for example insufficient quantity and inappropriate time.
- Corruption at management level.

Cost Coverage

The cost recovery is often insufficient for full cost coverage, for example:

Country	Cost recovery (%)	Cost coverage (%)	Remarks
Argentine	67	12	low tariff: $70/ha/year
Bangladesh	3-10	<1	tariffs not enforced
Brazil, Jaiba project	66	52	
Colombia	76	52	
Turkey	76	30-40	
Sri Lanka	8	<1	tariffs not enforced

Water User Associations (WUAs)

From 1980 programs were developed to transfer the operation and maintenance tasks from the government to water user associations (WUAs) that show some resemblance to water boards in the Netherlands, with the difference that it concerns irrigation rather than drainage and flood control. An effective development occurred in Mexico, where in 1990 a program of WUAs was initiated with tradable water rights. By 1998 some 400 WUAs were in operation commanding on average 7600 ha per WUA. They were able to recover more than 90% of the tariffs, mainly because they had to be paid in advance. Government subsidies to the water distribution and maintenance reduced to only 6%.

Water Delivery Principles

Rotational Turns

In large irrigation schemes, the distribution of irrigation water and the delivery at the farm gate is often arranged by *rotational turns* (e.g. every fortnight). The quantity of water to be received is often proportional to the farm size. As the canals usually transport constant flows, the water is being received during a period of time proportional to the farm size (e.g. every fortnight during 2 hours). The method of rotational turns is typical for the *utility form* of irrigation management.

Irrigation water delivered at the farm.

On Demand

In smaller irrigation schemes the water delivery may be arranged "on demand" with water charges are on a volumetric basis. This requires a precise bookkeeping system. As the demand may be fluctuating over time, the distribution system and infrastructure is relatively expensive because it must be able to cope with periods of peak demand. During periods of water scarcity, negotiations are due to regulate the supply or restriction agreements must have been made.

From point of view of efficient irrigation water-use this is the most effective system.

Preferential Rights

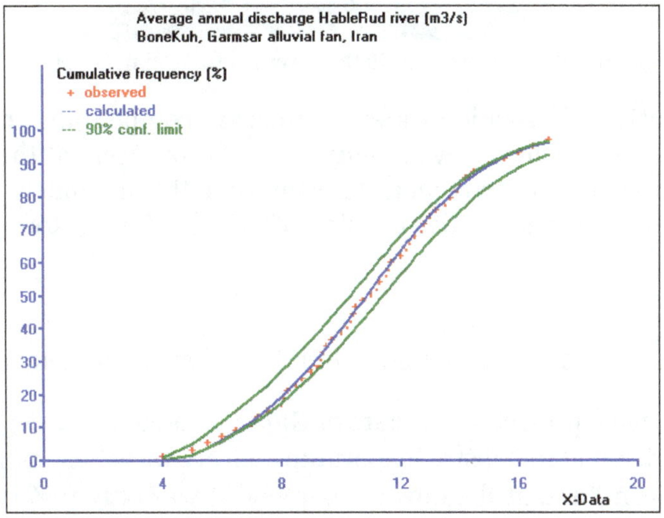

Cumulative frequency analysis of the variable annual discharge of a river.

In projects with an uncertain supply of water due to annual variations in river discharge, water users at the top (the head users) of the irrigation system (i.e. near the system's take-off point) often have preference, to a certain extent, over users at the tail-end. Hence, the number of farmers that are able to grow an irrigated crop may vary from year to year according to the riparian water rights.

The preferential method of irrigation can be found in spate irrigation systems. It is likely that the irrigators near the headworks, or their ancestors, did contribute more to the construction and maintenance of the works than the others, and therefore acquired the preferential rights.

Protective Duty

In regions with a structural water scarcity, the principle of *water duty* is often applied, whereby the duty per ha per season is only a fraction of the full irrigation need per ha (i.e. the *irrigation intensity* is less than 100%). Thus, farmers can irrigate only part of their land or irrigate their crops with a limited amount of water, whereby they may choose between crops with a high consumptive use (e.g. rice, sugarcane, most orchards) or a low consumptive use (e.g. cereals - notably barley, millet, and sorghum - or cotton). In India, such practice is called *protective irrigation*, which aims at equal distribution of scarce means and prevention of acute famine.

The method of protective irrigation is typical for the *utility form* of irrigation management.

Owing to competition for water, the water delivery practices may deviate from the principles.

Water Delivery Practices

Competition

The old diversion weir in the River Nile at Assiut, Egypt.

In practice the distribution of irrigation water is subject to competition. Influential farmers may be able to acquire more water than they are entitled to. Water users at the upstream part of the irrigation system can more easily intercept extra water than the tail-ender. The degree of farmers' influence is often correlated to the relative position of their land in the topography of the scheme.

Tail-end Problems

R.Chambers cites authors who have reported tail-end problems. Examples are:

- The old Sardar canal project in the state of Gujarat, India, was designed with an irrigation intensity of 32%, but at the upstream part the delivery was at an intensity of 42% (i.e. 131% of the design norm) and at the downstream end it was only 19% (i.e. 59% of the norm), although the project aimed at protective irrigation with equal rights for all.
- The Sardar Sahayak Pariyojana irrigation project, an extension of the Sardar canal project with 1.7 million ha, the head farmers received 5 times more water than the tail-enders, although the project was designed for equal distribution of the scarce water.
- The Ghatampur distributary canal in the Ramganga irrigation project in the state of Uttar Pradesh, India, delivered an amount of water equal to 155% of the design discharge to the Kisarwal district canal near the head of the distributary and only 22% to the Bairampur district canal at the downstream end.

The Ibrahimiya irrigation canal near Minya, Egypt.

Also in Egypt, in 1984, considerable differences in the water distribution over the canal systems have been reported:

Lateral canal	Water supply in m^3/feddan)
Kafret Nasser	4700
Beni Magdul	3500
El Mansuria	3300
El Hammami (upstream)	2800
El Hammami (downstream)	1800
El Shimi	1200

IRRIGATION SCHEDULING

Irrigation scheduling is the process used by irrigation system managers to determine the correct frequency and duration of watering.

The following factors may be taken into consideration:

- Precipitation rate of the irrigation equipment - How quickly the water is applied, often expressed in inches or mm per hour.
- Distribution uniformity of the irrigation system - How uniformly the water is applied, expressed as a percentage, the higher the number, the more uniform.
- Soil infiltration rate - How quickly the water is absorbed by the soil, the rate of which also decreases as the soil becomes wetter, also often expressed in inches or mm per hour.
- Slope (topography) of the land being irrigated as this affects how quickly runoff occurs, often expressed as a percentage, i.e. distance of fall divided by 100 units of horizontal distance (1 ft of fall per 100 ft (30 m) would be 1%).
- Soil available water capacity, expressed in units of water per unit of soil, i.e. inches of water per foot of soil.
- Effective rooting depth of the plants to be watered, which affects how much water can be stored in the soil and made available to the plants.
- Current watering requirements of the plant (which may be estimated by calculating evapotranspiration, or ET), often expressed in inches per day.
- Amount of time in which water or labor may be available for irrigation.
- Amount of allowable moisture stress which may be placed on the plant. For high value vegetable crops, this may mean no allowable stress, while for a lawn some stress would be allowable, since the goal would not be to maximize production, but merely to keep the lawn green and healthy.
- Timing to take advantage of projected rainfall.

- Timing to take advantage of favorable utility rates.
- Timing to avoid interfering with other activities such as sporting events, holidays, lawn maintenance, or crop harvesting.

The goal in irrigation scheduling is to apply enough water to fully wet the plant's root zone while minimizing overwatering and then allow the soil to dry out in between waterings, to allow air to enter the soil and encourage root development, but not so much that the plant is stressed beyond what is allowable.

In recent years, more sophisticated irrigation controllers have been developed that receive ET input from either a single on-site weather station or from a network of stations and automatically adjust the irrigation schedule accordingly.

Other devices helpful in irrigation scheduling are rain sensors, which automatically shut off an irrigation system when it rains, and soil moisture sensing devices such as capacitance sensors, tensiometers and gypsum blocks.

ADAPTIVE PARTICIPATORY INTEGRATED APPROACH

The adaptive participatory integrated approach (APIA) is a method of developing and managing water irrigation in developing regions. It attempts to balance multiple competing needs for water among various users.

The prevailing sectoral approach to irrigation management has tended to prevent the optimal use of water for agriculture and fisheries. Adoption of an adaptive, participatory and integrated assessment (APIA) of the impacts of irrigation on fisheries can ensure that poverty is alleviated, and food security and livelihoods are enhanced rather than hurt by irrigation development. It also provides an opportunity to decrease conflicts between fishers and farmers, and to increase overall benefits of irrigation systems at little additional cost.

The approach has been designed in interaction with a multidisciplinary team of researchers, as part the production of the "Guidance Manual for the Management of Impacts of Irrigation Development on Fisheries". The proposed APIA approach was tested through an ex-ante impact assessment in Lao PDR and an ex-post assessment in Sri Lanka.

In Laos, the ex-ante appraisal of the damming of a small river predicted a modest positive impact on production and livelihoods, with creation of a reservoir fishery outweighing degradation of the pre-existing river fishery. The outcome was contingent on continuation of rainfed wet season rice production and traditional farming practices, as adoption of practices involving reduced water storage in paddy fields was likely to reduce fish production considerably. Since maintaining connectivity of paddy fields and perennial water bodies is likewise important, cross-drainage culverts need to be passable to fish and the harvesting at these concentration points should be restricted. The creation of a reservoir would improve access to and convenience of fishing for those within reach, particularly in the dry season, and it would increase opportunities for specialisation in fishing. New income opportunities in irrigated agriculture are likely to reduce the overall level of fishing effort and increase returns, with the greatest benefits to the households most reliant on fishing.

In Sri Lanka, an ex-post evaluation of reservoir construction and the expansion of irrigated command areas revealed a complex picture of both positive and negative impacts at different points in the watershed. Over the catchment as a whole (including upper reaches, floodplain, coastal lagoons, and new and pre-existing reservoirs), the net effect on the value of aggregate production was estimated to have been positive. Sustaining this benefit was contingent on adoption of improved regimes for the management of reservoir water levels and fish stocks. Improvements in the management of the hydrological regime of coastal lagoons would also further raise aggregate production. Although practiced by a minority of rural households, fishing was found to perform a wide range of livelihood functions, including: livelihood of last resort (landless fishers); part of a diversified subsistence or accumulation strategy; or a specialist occupation. Disaggregated analysis of the impacts on the livelihoods of households who practiced fishing showed different effects, depending on where the household fished, and the functions performed by fishing in their livelihood strategy.

In both cases, irrigation development was deemed to have positive net impacts on local fisheries production and to provide livelihood opportunities related to fisheries, in particular to the poor. These gains and opportunities arise from specific local conditions, and can only be realized through the use of appropriate water management, agricultural practices and possibly fisheries regulations.

APIA facilitated the appraisal of gross production and livelihoods impacts, and opportunities for mitigation of negative or enhancement of positive impacts. Testing and systematic evaluation validated the approach and also revealed the following benefits and limitations.

The benefits of APIA compared to a conventional narrow technical assessment have been shown to include the following:

1. The ability to study inland fisheries in a larger context of multiple competing uses for water and alternative livelihood opportunities, with explicit identification of conflicts within a sequential, structured and management oriented process.
2. The aim for addressing the most important issues and solutions to be sought.
3. The emphasis on existing opportunities and the need for workable and cost effective mitigation and enhancement measures, with avoidance where possible of costly modifications to the irrigation infrastructure or other engineering solutions.
4. The strong motivation for stakeholders to seek out complementarities between irrigation and fisheries and to minimize trade-offs, with the potential to determine "win-win" solutions from resolution of conflicts between farmers and fishers.

However, use of APIA approach can be demanding of time, expertise for its implementation and in requirements for its support. In particular adequate awareness and strong institutional support are essential at all relevant levels of governance. Participatory processes can be biased if the representation and influence of competing groups is unbalanced, and there is risk that undue emphasis is given to conflicts that are difficult to resolve. Participation and local knowledge can be of great value, but should not be a substitute for rigorous technical assessment where this is necessary. Finally cumulative and synergistic impacts on fisheries of multiple irrigation schemes and other water resource developments within a river basin need to be further investigated.

IRRIGATION SYSTEM MAINTENANCE

Landscape irrigation uses 50 percent or more of the water provided by many utilities. A poorly maintained irrigation system means that much of this water never reaches its intended source and is lost to runoff, evaporation and deep watering below the root zone.

Maintaining irrigation systems is one most effective ways to reduce wasted water, reduce pollution from run-off and over-irrigation, and improve plant health by applying the correct amount of water where it can be utilized by the landscape.

At a minimum, a check of the irrigation system should be performed twice seasonally. Once at the beginning of the season when the system is first turned on and again halfway through the season. Ideally, the system should be checked on a monthly basis.

The basics of irrigation maintenance are:

1. Inspect the controller and make sure it's plugged in and functioning.
2. Update the time and date.
3. Check the connection on all of the wires – Make sure that rain, wind, or soil moisture sensors are connected.
4. Replace the back-up battery.
5. Change the schedule to reflect the current season and irrigation needs of the landscape.
6. Turn on each zone and look for system damage.

Examples of System Damage and Problems

Leaking Valves or Pipes

Leaks can occur as a result of weather damage (freezing and thawing), damage from shovels and other sharp tools, vandalism, tree roots, and normal aging of the system. Leaks from valves and pipes may be large and very obvious. Smaller leaks may not show up immediately and will require some detective work. Replace or repair damaged valves and pipes.

Broken or missing heads Damage can occur to sprinkler heads from lawn mowers, vandalism, improperly installed heads, and normal wear and tear. Replace damaged or missing heads immediately. Installing heads on swing pipe allows the head to "float" in the soil and reduces the damage that can result from lawn mowers and other heavy objects.

Clogged Nozzles

Clogged nozzles occur as a result of debris entering the irrigation system, a dirty water source, and normal wear and tear. Flush system at the beginning of the irrigation system, install screens on sprinkler heads, replace clogged nozzles, and improve system filtration.

Seal Leaks

Over time, dirt and debris can wear out the wiper seal resulting in leaks around the top of the spray head. If the spray head consists of a single unit the entire head must be replaced; with some spray heads it is possible to screw off the top of the sprinkler head and replace.

Sunken Heads

It is not uncommon for sprinkler heads to settle over time. Even when the soil is packed around them during, the weight of lawn mowers and other heavy equipment on wet turf can cause the heads to settle. Grass clipping, soil, and other debris can build up around heads resulting in a head that doesn't clear the grass adequately and disrupts the spray pattern. Current best management practices call for higher mowing heights – older systems may have been designed for shorter turf. Heads can be raised by using taller sprinkler bodies, or installing risers. Cutting turf away from heads is another solution but must be seasonally.

Tilted Heads

Lawn mowers and wet soil can cause newly installed sprinkler heads to tilt resulting in uneven coverage. If possible install the sprinkler head on swing pipe and move the head out of the line of mowers and other equipment that may cause the head to settle or shift. Otherwise reposition the head and pack the soil around it carefully.

Table shows an estimate of the number of gallons wasted annually from damage that can occur to the system. Water wasted (gallons) is calculated on a 26-week irrigation season, 3.5 events weekly (91 irrigation events) and 20 minutes per event.

Table: Water waste from typical irrigation system damage.

Problem	Flow Rate (gpm)	Run time/event (minutes)	Irrigation events/ week	Events/ Year	Waste/year (gal)
Lateral line leaks	5	20	3.5	91	9,100
Seal leak	4	20	3.5	91	7,280
Break in drip line	3	20	3.5	91	5,460
Missing or broken head	6	20	3.5	91	10,920

Problems affecting the Efficiency of the System

Misaligned Heads

Misaligned heads result in overwatering parts of the landscape and brown spots in other areas of the landscape. Realign heads according to manufacturer's instructions.

Overspray (Runoff)

Overspray can result in poor distribution uniformity, overwatering, and runoff on hardscapes. Reduce or eliminate overspray by adjusting system pressure, adjusting nozzles, or installing the proper size and type of nozzle.

Pressure Problems

High pressure causes problems such as misting from spray heads and rapid rotation of impact rotors resulting in poor coverage and potential damage to the irrigation system. Installation or adjustment of a pressure regulator or installation of pressure regulating heads can eliminate the problem.

Failure of heads to pop-up or rotate is often a symptom of low water pressure. Low pressure can also result in poor coverage and dry spots in the landscape. Adding a booster pump, watering during off-peak times, or reducing the number of heads on a zone can increase pressure.

Incorrect Spray Arc

Observe the spray pattern of the sprinkler heads on each zone and adjust the pattern if possible. If necessary, replace the nozzle with one that has the correct spray pattern. Older systems in particular may need to be updated with different nozzles as manufacturers are making nozzles to fit strips and narrow areas.

Table shows an estimate of the number of gallons wasted annually from system inefficiencies. Water wasted (gallons) is calculated on a 4,500 gallon (17.03 m^3) event over a 26-week irrigation season and 3.5 events weekly (91 irrigation events).

Table: Water waste from typical irrigation inefficiencies.

Problem	Estimate of event volume affected	Volume of total event (gal)	Events/ week	Events/ Year	Waste/year (gal)
Misting (high pressure)	10%	4,500	3.5	91	40,950
Misaligned heads	5%	4,500	3.5	91	20,475
Overspray	15%	4,500	3.5	91	61,425
Mismatched heads	4%	4,500	3.5	91	16,380

Although some of the repairs may require the help of a landscape professional many of them are relatively simple and can be easily mastered by the homeowner. Irrigation maintenance is one of the most important factors in reducing water waste in the landscape. In addition, a well-maintained and properly adjusted system is essential for a healthy, beautiful, landscape.

5

Irrigation Machinery

There are many important machines that are used in irrigation such as soil moisture sensors, irrigation controllers, irrigation system heads, disc filters, irrigation sprinklers and sprinkler system timers. The diverse applications of these irrigation machines have been thoroughly discussed in this chapter.

SOIL MOISTURE SENSOR

In the context of water management for irrigation, measuring and monitoring soil water status is an essential component of best management practices (BMPs) to improve the sustainability of agriculture.

Water content in the soil can be directly determined using the difference in weight before and after drying a soil sample. This direct technique is usually referred to as the thermo-gravimetric method (or simply gravimetric) when expressing water content as weight of water over weight of dry soil, GWC (i.e., the ratio of the mass of water present in a sample to the mass of the soil sample after it has been oven-dried (100-110 °C) to a constant weight). On the other hand, the thermo-volumetric method (or simply volumetric) gives the water content as volume of water in a volume of undisturbed soil VWC (i.e., volume of water related to the volume of an oven-dried undisturbed sample (soil core)). Although these direct methods are accurate and inexpensive, they are destructive, slow (2 days minimum), time-consuming and do not allow for making repetitions in the same location. Alternatively, many indirect methods are available for monitoring soil water content. These methods estimate soil moisture by a calibrated relationship with some other measurable variable. The suitability of each method depends on several issues like cost, accuracy, response time, installation, management and durability.

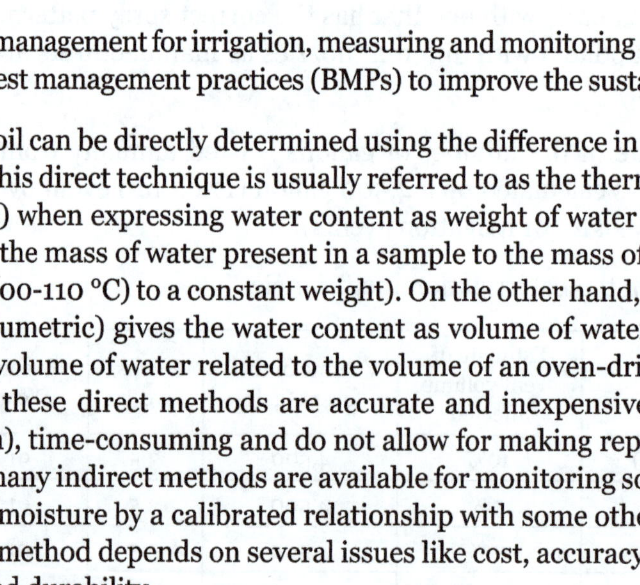

Methods for soil moisture measuring.

Depending on the quantity measured, indirect techniques are first classified into volumetric and tensiometric methods. While the former gives volumetric soil moisture, the latter yields soil suction or water potential (i.e., tension exerted by capillarity). Both quantities are related through the soil water characteristic curve specific to a given soil.

It is important to remember that each soil type (texture/structure) has a different curve; therefore, they cannot be related to each other the same way for all soil types. In addition, this relationship might not be unique and may differ along drying and wetting cycles, especially in finer soils. Several mathematical equations have been proposed to describe the soil characteristic curve that can help to calculate water needed for irrigation.

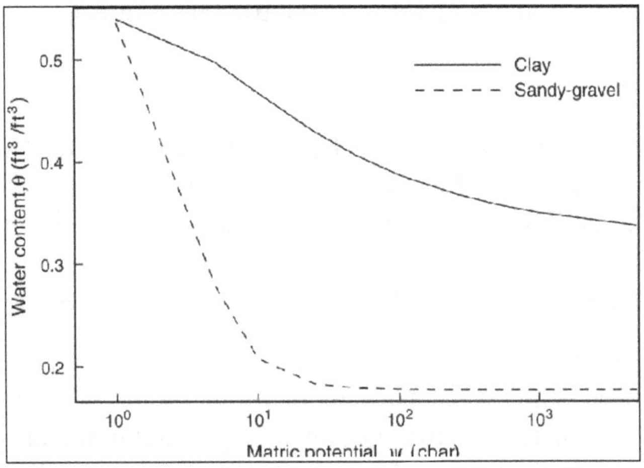

Measured soil characteristic curves for two different soil types.

Depending on the soil physical properties and goal of the soil moisture measurement, some devices are more effective than others. Firstly, it must be considered that although volumetric moisture is a more intuitive quantity, in fine texture soils water is strongly retained by solid particles and therefore may not be available for plant absorption and other processes like flow. In the case of plant-soil studies, soil suction may be a more useful quantity since it relates to the energy that plants have to invest to extract the water from the soil, and hence it is a more meaningful measure of plant water stress. Secondly, the desired sampling frequency is an important factor since response times of different sensors vary over a wide range (i.e., some devices require soil moisture to equilibrate with the sensor matrix). Thirdly, soil physical properties (texture, shrinking/swelling) may influence the suitability of the selected method, because some of them require good soil-instrument contact. On the other hand, depending on soil type and hydrologic conditions (precipitation and evapotranspiration), some instruments might have higher maintenance requirements than others.

Irrigation management is a practical application of monitoring soil moisture that is becoming widespread among agricultural growers. Soil moisture-based optimized irrigation consists of keeping the soil within a target moisture range by replenishing the plant water uptake with irrigation. This practice reduces the potential for soil water excess and leaching of agrochemicals present in the soil, however it requires selection of a suitable method for soil moisture estimation. To calculate irrigation requirements (the amount of water that needs to be applied with each irrigation), suction values from tensiometric methods need to be converted to soil moisture through the soil characteristic curve.

Field Methods for Soil Moisture Measurement: Trade-offs

Most practical techniques for soil water monitoring are indirect.

Volumetric Field Methods

All methods under this definition estimate the volume of water in a sample volume of undisturbed soil. This quantity is useful for determining how saturated the soil is (i.e., fraction of total soil volume filled with the soil aqueous solution). When it is expressed in terms of depth [i.e., volume of water in soil down to a given depth over a unit surface area (inches of water)], it can be compared with other hydrological variables like precipitation, evaporation, transpiration, deep drainage, etc.

Neutron Moderation

Working principle: Fast neutrons are emitted from a decaying radioactive source (241Am/9Be) and when they collide with particles having the same mass as a neutron (i.e., protons, H^+), they slow down dramatically, building a "cloud" of "thermalized" (slowed-down) neutrons. Since water is the main source of hydrogen in most soils, the density of slowed-down neutrons formed around the probe is nearly proportional to the volume fraction of water present in the soil.

Description: The probe configuration is in the form of a long and narrow cylinder, containing a source and detector. Measurements are made by introducing the probe into an access tube. It is possible to determine soil moisture at different depths by hanging the probe in the tube at different depths. The soil moisture is obtained from the device based on a linear calibration between the count rate of slowed-down neutrons at the field, and the soil moisture content obtained from nearby field samples.

Neutron probe.

Advantages

- Robust and accurate.
- Inexpensive per location (i.e., a large number of measurements can be made at different points with the same instrument).
- One probe allows for measuring at different soil depths.
- Large soil sensing volume (sphere of influence with 4-16 in. (10.2 cm - 40.6 cm) radius, depending on moisture content).

- Not affected by salinity or air gaps.
- Stable soil-specific calibration.

Drawbacks

- Safety hazard, since it implies working with radiation. Even at 16 in (40.6 cm) depth, radiation losses through soil surface have been detected.
- Requires certified personnel.
- Requires soil-specific calibration.
- Heavy, cumbersome instrument.
- Takes relative long time for each reading.
- Readings close to the soil surface are difficult and not accurate.
- Manual readings; cannot be automated due to hazard.
- Expensive to buy.

The sphere of influence may vary according to the following reasons:

- It increases as the soil dries, because the hydrogen concentration reduces, so that the probability of collision is smaller and thereby fast neutrons can travel further from the source.
- It is smaller in fine texture soils, because they can hold more water, thus the probability of collision is higher.
- If there are layers with large differences in water content due to changes in soil physical properties, the sphere of influence can have a distorted shape.

Dielectric Methods

The next set of volumetric methods is known as dielectric techniques. They estimate soil water content by measuring the soil bulk permittivity (or dielectric constant), Kab, that determines the velocity of an electromagnetic wave or pulse through the soil. In a composite material like the soil (i.e., made up of different components like minerals, air and water), the value of the permittivity is made up by the relative contribution of each of the components. Since the dielectric constant of liquid water (Kaw = 81) is much larger than that of the other soil constituents (e.g. Kas = 2-5 for soil minerals and 1 for air), the total permittivity of the soil or bulk permittivity is mainly governed by the presence of liquid water.

A common approach to establish the relationship between Kab and volumetric soil moisture (VWC) is the empirical equation of Topp et al:

$$VWC = -5.3 \cdot 10^{-2} + 2.29 \cdot 10^{-2} Kab - 5.5 \cdot 10^{-4} Kab^{2} + 4.3 \cdot 10^{-6} Kab^{3}$$

This relationship works for most mineral soils (independent of soil composition and texture) and for moisture below 50%. For larger water content, organic or volcanic soils, a specific calibration is required. It is worth noticing that the relationship depends on the electromagnetic wave frequency sent by the specific device. At low frequencies (<100 MHz) it is more soil-specific.

The dielectric methods described below use empirical calibrated relationships between VWC and the sensor output signal (time, frequency, impedance, wave phase). These techniques are becoming widely adopted because they have good response time (almost instantaneous measurements), do not require maintenance, and can provide continuous readings through automation.

Time Domain Reflectometry (TDR)

1. Working principle: The soil bulk dielectric constant (Kab) is determined by measuring the time it takes for an electromagnetic pulse (wave) to propagate along a transmission line (TL) that is surrounded by the soil. Since the propagation velocity (v) is a function of Kab, the latter is therefore proportional to the square of the transit time (t, in seconds) down and back along the TL:

$$Kab = (c/v)^2 = ((c.t)/(2.L))^2$$

Where, c is the velocity of electromagnetic waves in a vacuum (3•108 m/s or 186,282 mile/s) and L is the length of the TL embedded in the soil (in m or ft).

2. Description: A TDR instrument requires a device capable of producing a series of precisely timed electrical pulses with a wide range of high frequencies used by different devices (e.g., 0.02-3 GHz), which travel along a TL that is built with a coaxial cable and a probe. This high frequency provides a response less dependent on soil specific properties like texture, salinity or temperature. The TDR probe usually consists of 2-3 parallel metal rods that are inserted into the soil acting as waveguides in a similar way as an antenna used for television reception. At the same time, the TDR instrument uses a device for measuring and digitizing the energy (voltage) level of the TL at intervals down to around 100 picoseconds. When the electromagnetic pulse traveling along the TL finds a discontinuity (i.e., probe-waveguides surrounded by soil) part of the pulse is reflected. This produces a change in the energy level of the TL. Thereby the travel time (t) is determined by analyzing the digitized energy levels.

Soil salinity or highly conductive heavy clay contents may affect TDR, since it contributes to attenuation of the reflected pulses. In other words, TDR is relatively insensitive to salinity as long as a useful pulse is reflected (i.e., as long as it can be analyzed). In soils with highly saline conditions, using epoxy-coated probe rods should solve the problem. However, this implies loss of sensitivity and change in calibration. It is interesting to notice that in addition to time of travel another characteristic of the pulse traveling through the soil (i.e., change in size or attenuation of the pulse) can be related to the soil electrical conductivity. Based on this some commercial devices incorporate the possibility of measuring water content and soil salinity simultaneously.

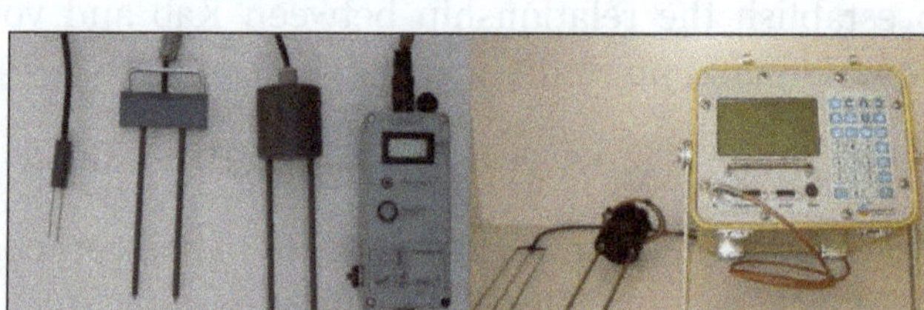

TDR equipment.

Advantages

- Accurate.
- Soil specific-calibration is usually not required.
- Easily expanded by multiplexing.
- Wide variety of probe configurations.
- Minimal soil disturbance.
- Relatively insensitive to normal salinity levels.
- Can provide simultaneous measurements of soil electrical conductivity.

Drawbacks

- Relatively expensive equipment due to complex electronics.
- Potentially limited applicability under highly saline conditions or in highly conductive heavy clay soils.
- Soil-specific calibration required for soils having large amounts of bound water (i.e., those with high organic matter content, volcanic soils, etc).
- Relatively small sensing volume (about 1.2 inch (3.05 cm) radius around length of wave-guides).

Frequency Domain (FD): Capacitance and FDR

Working principle: The electrical capacitance of a capacitor that uses the soil as a dielectric depends on the soil water content. When connecting this capacitor (made of metal plates or rods imbedded in the soil) together with an oscillator to form an electrical circuit, changes in soil moisture can be detected by changes in the circuit operating frequency. This is the basis of the Frequency Domain (FD) technique used in Capacitance and Frequency Domain Reflectometry (FDR) sensors. In Capacitance sensors the dielectric permittivity of a medium is determined by measuring the charge time of a capacitor made with that medium. In FDR the oscillator frequency is swept under control within a certain frequency range to find the resonant frequency (at which the amplitude is greatest), which is a measure of water content in the soil.

Description: Probes usually consist of two or more electrodes (i.e., plates, rods, or metal rings around a cylinder) that are inserted into the soil. On the ring configuration the probe is introduced into a access tube installed in the field. Thus, when an electrical field is applied, the soil around the electrodes (or around the tube) forms the dielectric of the capacitor that completes the oscillating circuit. The use of an access tube allows for multiple sensors to take measurements at different depths.

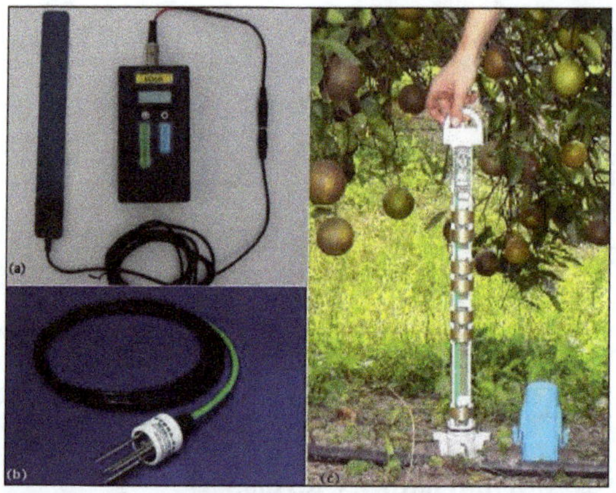

FD probes: a) Capacitance (plates imbibed in a silicon board);
b) Capacitance (rods); and c) FDR (rings).

A soil-specific calibration is recommended because the operating frequency of these devices is generally below 100 MHz. At these low frequencies the bulk permittivity of soil minerals may change and the estimation is more affected by temperature, salinity, bulk density and clay content.

Advantages

- Accurate after soil-specific calibration.
- Can read in high salinity levels, where TDR fails.
- Better resolution than TDR (avoids the noise that is implied in the waveform analysis performed by TDRs).
- Can be connected to conventional loggers (DC output signal).
- Flexibility in probe design (more than TDR).
- Some devices are relatively inexpensive compared to TDR due to use of low frequency standard circuitry.

Drawbacks

- The sensing sphere of influence is relatively small (about 1.6 in. (4.1 cm)).
- For reliable measurements, it is extremely critical to have good contact between the sensor (or tube) and soil.
- Careful installation is necessary to avoid air gaps.
- Tends to have larger sensitivity to temperature, bulk density, clay content and air gaps than TDR.
- Needs soil-specific calibration.

Amplitude Domain Reflectometry (ADR): Impedance

Working principle: When an electromagnetic wave (energy) travelling along a transmission line

(TL) reaches a section with different impedance (which has two components: electrical conductivity and dielectric constant), part of the energy transmitted is reflected back into the transmitter. The reflected wave interacts with the incident wave producing a voltage standing wave along the TL, i.e., change of wave amplitude along the length of the TL. If the soil/probe combination is the cause for the impedance change in the TL, measuring the amplitude difference will give the impedance of the probe. The influence of the soil electrical conductivity is minimized by choosing a signal frequency, so that the soil water content can be estimated from the soil/probe impedance.

Description: Impedance sensors use an oscillator to generate a sinusoidal signal (electromagnetic wave at a fixed frequency, e.g., 100 MHz) that is applied to a coaxial TL that extends into the soil through an array of parallel metal rods, the outer of which forms an electrical shield around the central signal rod. This rod arrangement acts as an additional section of the TL, having impedance that depends on the dielectric constant of the soil between the rods.

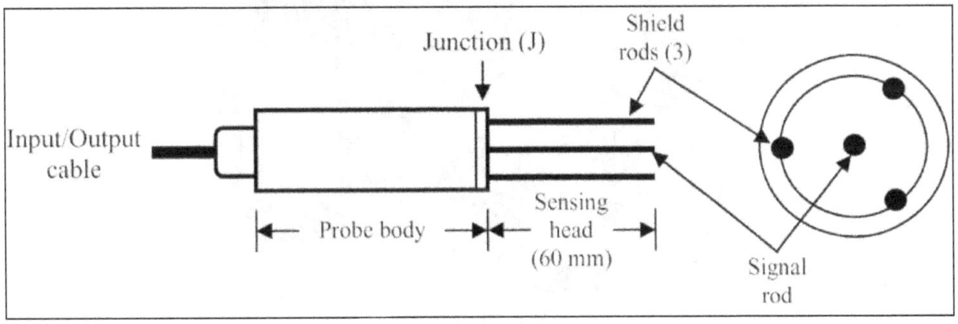

ADR probe4.

Advantages

- Accurate with soil-specific calibration.
- Allows measurements in highly saline conditions (up to 20 dS/m).
- Minimal soil disturbance.
- Can be connected to conventional loggers (DC output signal).
- Inexpensive due to standard circuitry.
- Not affected by temperature.
- *In situ* estimation of soil bulk density possible.

Drawbacks

- Soil-specific calibration recommended for reliable measurements.
- Measurement affected by air gaps, stones or channeling water directly onto the probe rods.
- Small sensing volume (0.27 in3 (4.42 ml)).

Phase Transmission (Virrib)

Working principle: After having travelled a fixed distance, a sinusoidal wave will show a phase shift relative to the phase at the origin. This phase shift depends on the length of travel along the TL, the frequency and the velocity of propagation. Since velocity of propagation is related to soil moisture content, for a fixed frequency and length of travel soil water content can be determined by this phase shift.

Description: The probe uses a particular waveguide design (two open concentric metal rings), so that phase measuring electronics can be applied at the beginning and ending of the waveguides.

Phase transmission probe.

Advantages

- Accurate with soil-specific calibration.
- Large sensing soil volume (4-5 gallons (15.2 L to 18.9 L)).
- Can be connected to conventional loggers (DC output signal).
- Inexpensive.

Drawbacks

- Significant soil disturbance during installation due to concentric rings sensor configuration.
- Requires soil-specific calibration.
- Sensitive to salinity levels >3 dS/m.
- Reduced precision, because the generated pulse gets distorted during transmission.
- Needs to be permanently installed in the field.

Time Domain Transmission (TDT)

Working principle: This method measures the one-way time for an electromagnetic pulse to propagate along a transmission line (TL). Thus, it is similar to TDR, but requires an electrical connection at the beginning and ending of the TL. Notwithstanding, the circuit is simple compared with TDR instruments.

Description: The probe has a waveguide design (bent metal rods), so that the beginning and ending of the transmission line are inserted into the electronic block. Alternatively, the sensor consists of a long band (~3 ft (91.4 cm)), having an electronic block at both ends.

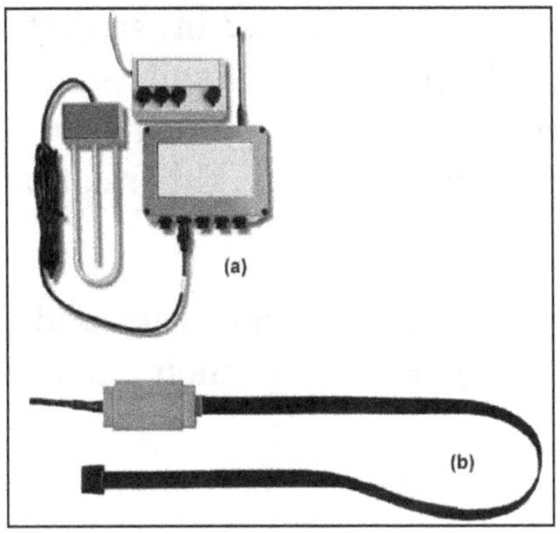

TDT probe.

Advantages

- Accurate.
- Large sensing soil volume (0.2-1.6 gallons (.76 L to 6.1 L)).
- Can be connected to conventional loggers (DC output signal).
- Inexpensive due to standard circuitry.

Drawbacks

- Reduced precision, because the generated pulse is distorted during transmission.
- Soil disturbance during installation.
- Needs to be permanently installed in the field.

Other Volumetric Field Methods

Another interesting technique is the Ground Penetrating Radar (GPR). This technique is based on the same principle as TDR, but does not require direct contact between the sensor and the soil. When mounted on a vehicle or trolley close to the soil surface, it has the potential of providing rapid, non-disturbing, soil moisture measurements over relatively large areas, whereas TDR is better for detailed measurements over small areas. Although it has been applied successfully to many field situations, GPR has not been widely used because the methodology and instrumentation are still only in the research and development phase. It is however likely that small, compact and inexpensive GPR systems will be available in the near future for routine field studies.

Mention should also be made of new remote sensing (non-contact) methods specially suited for soil moisture monitoring over large areas, usually mounted on airplanes or satellites. Among these methods, the active and passive microwave, and electromagnetic induction (EMI) methods have been found useful in different applications, and are the subject of much current research. The active and EMI methods use two antennae to transmit and receive electromagnetic signals that are reflected by the soil, whereas the passive microwave just receives signals naturally emitted by the soil surface. On the microwave methods, typically the signal relates to some shallow depth below the ground surface (<4 in) so that only a measure of the soil moisture and electrical conductivity of the near-surface soil can be achieved. EMI does not measure water content directly, but rather soil electrical conductivity, and a known calibration relationship between the two is required. Unfortunately, this relationship is site-specific and cannot be assumed.

Other modern techniques for estimating soil moisture and flow are x-ray tomography and nuclear magnetic resonance (NMR). However, their application to field conditions is lim ited.

Tensiometric Field Methods

Tensiometric methods estimate the soil water matric potential that includes both adsorption and capillary effects of the soil. The matric potential is one of the components of the total soil water potential that also includes gravitational (position with respect to a reference elevation plane), osmotic (salts in soil solution), gas pressure or pneumatic (from entrapped air), and overburden components. The sum of matric and gravitational potentials is the main driving force for water movement in soils and other soil-like porous media.

All available tensiometric instruments have a porous material in contact with the soil, through which water can move. Thereby, water is drawn out of the porous medium in a dry soil and from the soil into the medium in a wet soil. It is worth noticing, that in general, they do not need a soil specific calibration, however, in most cases they have to be permanently installed in the field, or a sufficiently long time must be allowed for equilibration between the device and the soil before making a reading.

Tensiometer

Working principle: When a sealed water-filled tube is placed in contact with the soil through a permeable and saturated porous material, water (inside the tube) comes into equilibrium with the soil solution (i.e., it is at the same pressure potential as the water held in the soil matrix). Hence, the soil water matric potential is equivalent to the vacuum or suction created inside the tube.

Description: The tensiometer consists of a sealed water-filled plastic tube with a ceramic cup at one end and a negative pressure gauge (vacuometer) at the other. The shape and size of the ceramic cup can be variable and the accuracy depends on the gauge or transducer used (about 0.01 bar). Typically the measurement range is 0-0.80 bar, although there are low-tension versions (0-0.40 bar) designed for coarse soils.

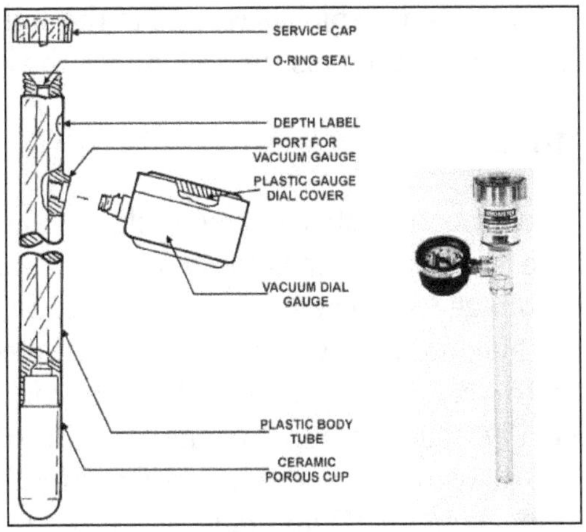

Tensiometer.

Advantages

- Direct reading.
- Up to 4 inch (10.2 cm) measurement sphere radius.
- Continuous reading possible when using pressure transducer.
- Electronics and power consumption avoidable.
- Well-suited for high frequency sampling or irrigation schedules.
- Minimal skill required for maintenance.
- Not affected by soil salinity, because salts can move freely in and out across the porous ceramic cup.
- Inexpensive.

Drawbacks

- Limited soil suction range (<1 bar).
- Relatively slow response time.
- Requires intimate contact with soil around the ceramic cup for consistent readings and to avoid frequent discharge (breaking of water column inside).
- Especially in swelling or coarse soils, the ceramic cup can loose contact with soil, thus requiring reinstallation.
- Requires frequent maintenance (refilling) to keep the tube full of water, specially in hot dry weather.

Resistance Blocks

Working principle: The electrical resistance between electrodes embedded in a porous medium (block) is proportional to its water content, which is related to the soil water matric potential of the surrounding soil. Electrical resistance reduces as the soil, hence the block, dries.

Gypsum (Bouyoucos) Block

Description: A gypsum block sensor constitutes an electrochemical cell with a saturated solution of calcium sulphate as electrolyte. The resistance between the block-embedded electrodes is determined by applying a small AC voltage (to prevent block polarization) using a Wheatstone bridge. Since changes to the soil electrical conductivity would affect readings, gypsum is used as a buffer against soil salinity changes (up to a certain level). The inherent problem is that the block dissolves and degrades over time (especially in saline soils) losing its calibration properties. It is recommended that the block pore size distribution match the soil texture being used. The readings are temperature dependent (up to 3% change/°C) and field measured resistance should be corrected for differences between calibration and field temperatures. Some reading devices contain manual or self-compensating features for temperature or the manufacture provides correction charts or equations. Measurement range is 0.3-2.0 bar.

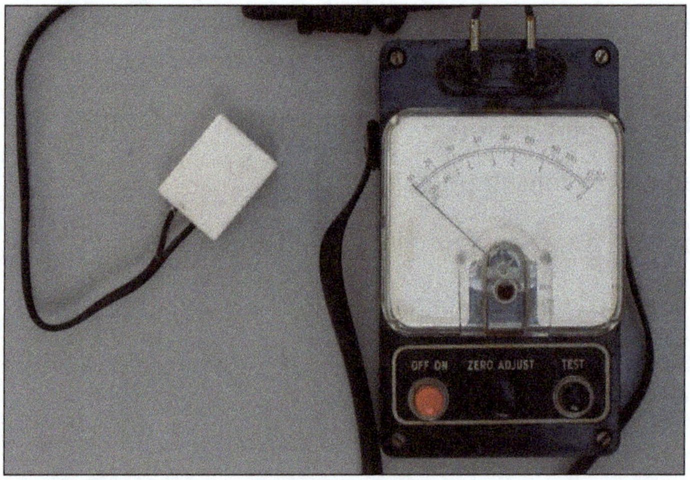

Gypsum block.

Advantages

- Up to 4 inch (10.2 cm) measurement cylinder radius.
- No maintenance needed.
- Simple and inexpensive.
- Salinity effects buffered up to 6 dS/m.
- Well suited for irrigation where only "full" and "refill" points are required.
- Suited to regulated-deficit irrigation.

Drawbacks

- Low resolution, limited use in research.

- Block cannot be used for measurements around saturation (0-0.3 bar).
- Block properties change with time, because of clay deposition and gypsum dissolution. Degradation speed depends on soil type, amount of rainfall and irrigation, and also the type of gypsum block used.
- Very slow reaction time. It does not work well in sandy soils, where water drains more quickly than the instrument can equilibrate.
- Not suitable for swelling soils.
- Inaccurate readings due to the block hysteresis (i.e., at a fixed soil water potential, the sensor can display different resistance when wetting than when drying).
- Temperature dependent. If connected to a logging system, another variable and sensor for temperature must be added to the system.

Granular Matrix Sensors (GMS)

Description: The sensor consists of electrodes embedded in a granular quartz material, surrounded by a synthetic membrane and a protective stainless steel mesh. Inside, gypsum is used to buffer against salinity effects. This kind of porous medium allows for measuring in wetter soil conditions and lasts longer than the gypsum blocks. However, even with good sensor-soil contact, GMS have rewetting problems after they have been dried to very dry levels. This is because of the reduced ability of water films to re-enter the coarse medium of the GMS from a fine soil. The GMS material allows for measurements closer to saturation. Measurement range is 0.10-2.0 bar.

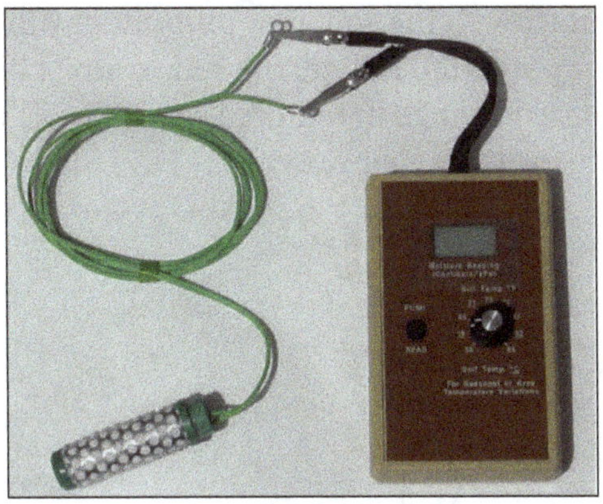

GMS sensor.

Advantages

- Reduces the problems inherent to gypsum blocks (i.e., loss of contact with the soil by dissolving, and inconsistent pore size distribution).
- Up to 4 inch (10.2 cm) measurement cylinder radius.
- No maintenance needed.
- Simple and inexpensive.

- Salinity effects buffered up to 6 dS/m.
- Suited to regulated-deficit irrigation.

Drawbacks

- Low resolution, limited use in research.
- Slow reaction time. It does not work well in sandy soils, where water drains more quickly than the instrument can equilibrate.
- Not suitable for swelling soils.
- If the soil becomes too dry, the sensor must be pulled out, re-saturated and installed again.
- Temperature dependence. If connected to a logging system, another variable and sensor for temperature must be added to the system.

Heat Dissipation

Working principle: The thermal conductivity of water produces heat dissipation, so that a dry material will heat up faster than a wet one. In other words, the heat flow in a porous material is proportional to its water content.

Description: A thermal heat probe consists of a porous block containing a heat source and an accurate temperature sensor. The block temperature is measured before and after the heater is powered for a few seconds. Thereby, block moisture is obtained from the temperature variation. Since the porous block, placed in contact with the soil, is equilibrated with the soil water, its characteristic curve will give the soil water potential. Hence, the sensor must be provided with the calibrated relationship between the measured change in temperature and soil water potential. Measurement range: 0.1-30 bar (less accurate for 10-30 bar range).

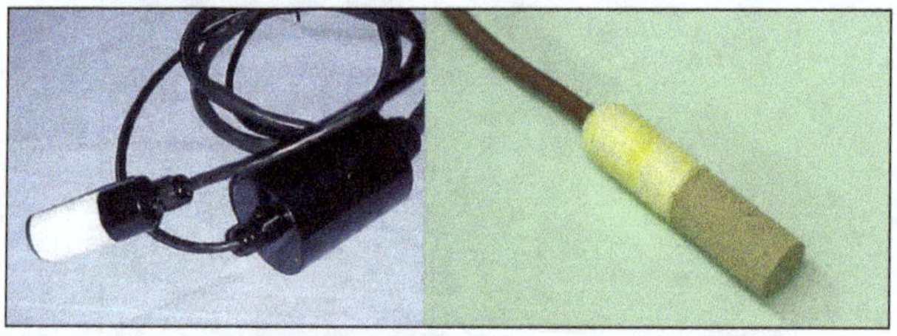

Heat dissipation sensor.

Advantages

- Wide measurement range.
- No maintenance required.
- Up to 4 inch (10.2 cm) measurement cylinder radius.
- Continuous reading possible.
- Not affected by salinity because measurements are based on thermal conductivity.

Drawbacks

- Needs a sophisticated controller/logger to control heating and measurement operations.
- Slow reaction time. It does not work well in sandy soils, where water drains more quickly than the instrument can equilibrate.
- Fairly large power consumption for frequent readings.

Soil Psychrometer

Working principle: Under vapour equilibrium conditions, water potential of a porous material is directly related to the vapour pressure of the air surrounding the porous medium. This means that the soil water potential is determined by measuring the RH of a chamber inside a porous cup equilibrated with the soil solution.

Description: A soil psychrometer consists of a ceramic shield or screen building an air chamber, where a thermocouple is located. The screen type is recommended for high salinity environments. RH in the air chamber is calculated from the "wet bulb" vs "dry bulb" temperature difference. Measurement range: 0.5-30 bar (less accurate for 10-30 bar range).

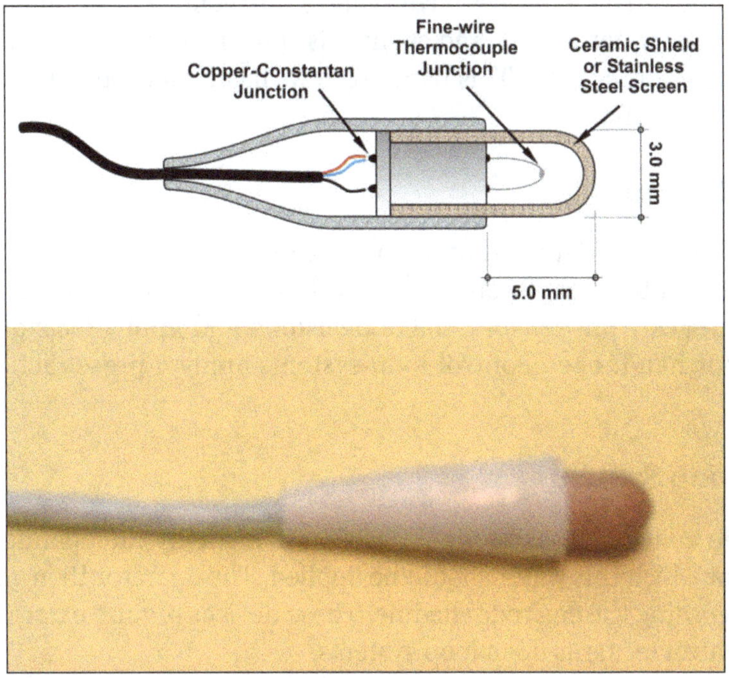

Soil psychrometer.

Advantages

- High sensitivity.
- Scientifically rigorous readings (except in wetter soil conditions).
- Suitable where typical moisture conditions are very dry.

Drawbacks

- Not recommended at shallow soil depths, due to high susceptibility to thermal gradient.
- Small sensing volume.
- Very slow reaction time, because reaching vapour equilibrium takes time.
- Low accuracy in the wet range.
- Specialized equipment is required for the sensor's excitation and reading.

IRRIGATION CONTROLLERS

A controller is an integral part of an irrigation system. It is an essential tool to apply water in the necessary quantity and at the right time to sustain agricultural production and to achieve high levels of efficiency in water, energy and chemical uses.

Irrigation controllers have been available for many years in the form of mechanical and electromechanical irrigation timers. These devices have evolved into complex computer-based systems that allow accurate control of water, energy and chemicals while responding to environmental changes and development stages of the crop. This paper reviews basic concepts related to control systems for irrigation management purposes.

Basic Control Strategies

Two general types of controllers are used to control irrigation systems: Open control loop systems, and closed control loop systems. The difference between these is that closed control loops have feedback from sensors, make decisions and apply decisions to the irrigation system. On the other hand, open control loop systems apply a preset action, as is done with irrigation timers.

Open Control Loop Systems

When using an open control loop system, a decision is made by the operator or the amount of water and the time at which this water should be applied. The operator then goes on to set an irrigation controller according the desired schedule. These devices require external intervention they are referred to in control terms as open loop systems.

Open loop control systems use irrigation duration or applied volume for control purposes. Figure shows the basic components of an open loop time-based irrigation controller. Notice that in this type of controller the basic control parameters are how often and how long is irrigation water is to be applied. Open loop controllers are also constructed in such a way that a clock is used to start irrigation and the application of a given volume to stop irrigation. In this type of controller the parameters set by the system operator are how often and the volume of water to be applied.

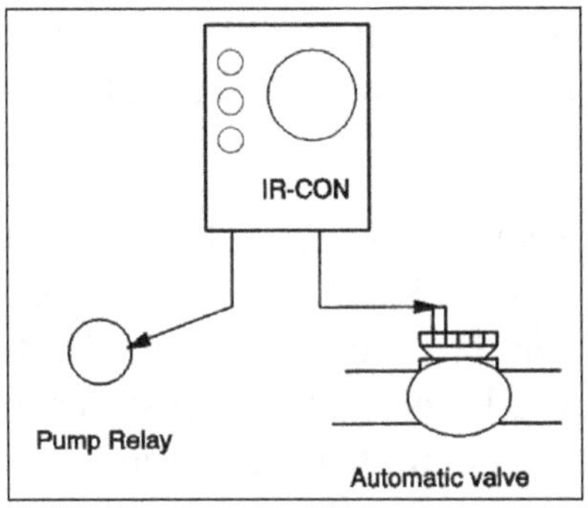

Open loop control systems have the advantages that they are low cost, readily available, and many variations of the devices are manufactured with different degrees of flexibility related to the number of stations and schedule specification. However, they do not respond automatically to changing conditions in the environment and require frequent resetting to achieve high levels of irrigation efficiency.

Closed Control Loop Systems

In a closed control loop the operator sets up a general strategy for control. Once the general strategy is defined, the control system takes over and makes detailed decisions of when to apply water and how much water to apply. This type of system requires that feedback be given back to the controller by one or more sensors. Depending on the feedback of the sensors, the irrigation decisions are made and actions are carried out if necessary. It is important to note that in this type of systems the feedback and control of the system is done continuously. Figure shows the elementary components of this type of system.

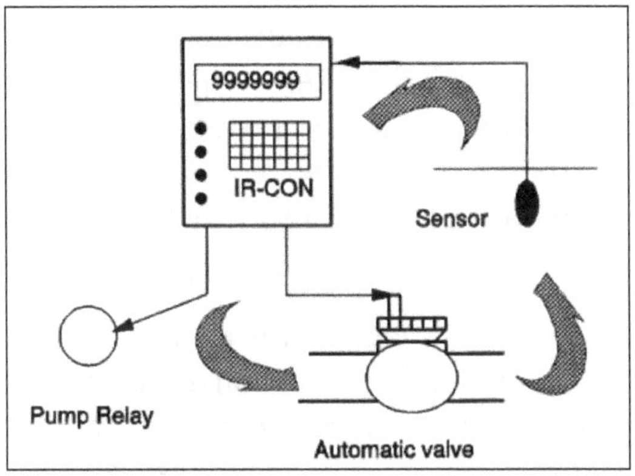

Closed loop controllers require data acquisition of environmental parameters, such as, soil-moisture, temperature, radiation, wind-speed and relative humidity. The state of the system (for example measured soil-moisture using a sensor as illustrated in figure) is compared against a desired state and a decision based on this comparison is made whether irrigation should be applied or not. Closed loop controllers for irrigation systems base their irrigation decisions on: 1) direct measurement of soil-moisture using sensors, 2) calculations of water used by the plants based on climatic parameters, or 3) both soil moisture sensors and climatic parameter measurements.

When using a computer-based controller, a very important component of a closed loop control system is the logic that is used to make decisions about operation of the irrigation system. Some of these systems may be very elaborate and use complicated simulation models that are verified with soil moisture measurements to arrive at an irrigation decision and implement the action at the appropriate time. Systems of this type (with different levels of complexity), are quickly being developed, and some have become commercially available in the past few years.

The simplest form of a closed loop control system is that of a high frequency irrigation controller that is interrupted by a moisture sensor. Figure shows this system. The sensor in figure is wired into the line that supplies power from the controller to the electric solenoid valve. The sensor operates as a switch that responds to soil moisture. When sufficient soil-moisture is available in the soil, the sensor maintains the circuit open. When soil-moisture drops below a certain threshold, the sensing device closes the circuit, allowing the controller to power the electrical valve.

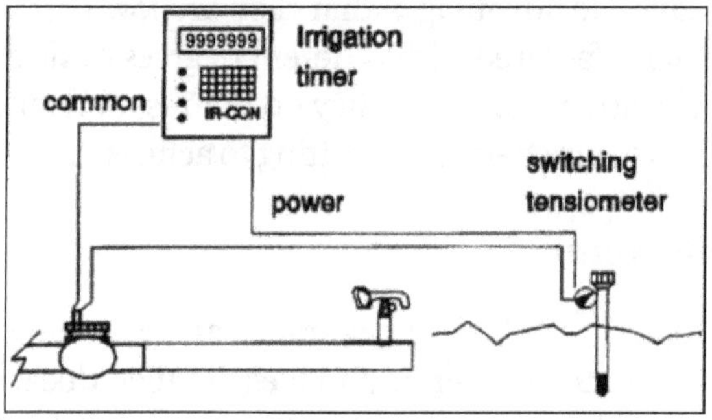

Using their arrangement in figure, the controller can be set to irrigate at a very high frequency (4 or 5 times more often than required). When the controller attempts to irrigate, irrigation will occur only if the soil-moisture sensor allows it, which in turn occurs only when soil-moisture has dropped below acceptable levels.

The system has been used successfully in controlling small sprinkler irrigated turf and microirrigated citrus at a research site using switching tensiometers. For turf, the tensiometers were installed at the center of the bottom third of the root system (10 inches deep) and the threshold was set to the point at which water stress symptoms were visible. In citrus a bank of sensors was used under the emitter connected in parallel, in such a way that any of the sensors would allow irrigation to occur.

The feedback system in figure is very low cost and is easy to install and maintain. However, the system has limitations: 1) Determining the best location of the sensor is not a straightforward task and requires some knowledge of soil-water and root dynamics, 2) spatial variability of soil properties may result in readings that are not representative of the system.

Irrigation Timers

Irrigation timers are simple controllers consisting of clock units capable of activating one or more subunits of the irrigation system at specified times. Several designs are commercially available with many different features and over a wide range of costs.

By today's standards most irrigation timers provide several of the following functions:

- A clock/timer: Provides the basic time measurements by which schedules are executed.
- A calendar selector: This function allows definition of which days the system is to operate.

- Station time setting: This function allows definition of start time and duration for each station.
- Manual start: This function allows the operator to start the automatic cycle without disturbing the preset starting time.
- Manual operation of each station: This allows the operator to manually start the irrigation cycle without making changes to the preset starting time.
- Master switch: This function prevents activation of any station connected to the timer.
- Station omission: This function is used to omit any specified number of stations from the next irrigation cycle.
- Master valve control: Provides control to a master system valve. This function is used with certain types of backflow prevention equipment and also prevents flow to the system in case of failure in the system.
- Pump start lead: This feature allows a pump start solenoid to be activated whenever a station is activated, thus tying pump control with irrigation control.

The two most common types of controller designs are electromechanical and electronic.

Electromechanical Controllers

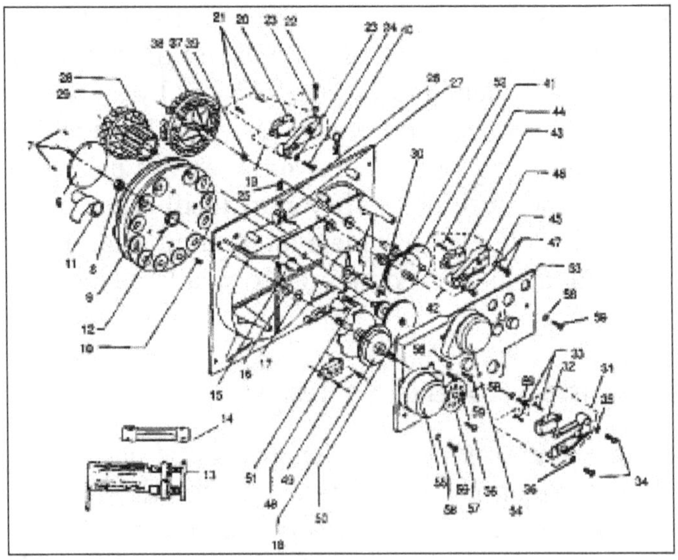

Electromechanical controllers use an electrically driven clock and mechanical switching (gear arrays) to activate the irrigation stations. These types of controllers are generally very reliable and not too sensitive to the quality of the power available. They generally are not affected by spikes in the power, and unless surges and brownouts are of such magnitude that they will damage the motor, they will continue to operate. Even if there is a power outage, the programmed schedule will not be lost and is generally delayed only for the duration of the power outage. However, because of the mechanically-based components they are limited in the features they provide. Figure shows the components of a commercially available electromechanical controller.

Electronic Controllers

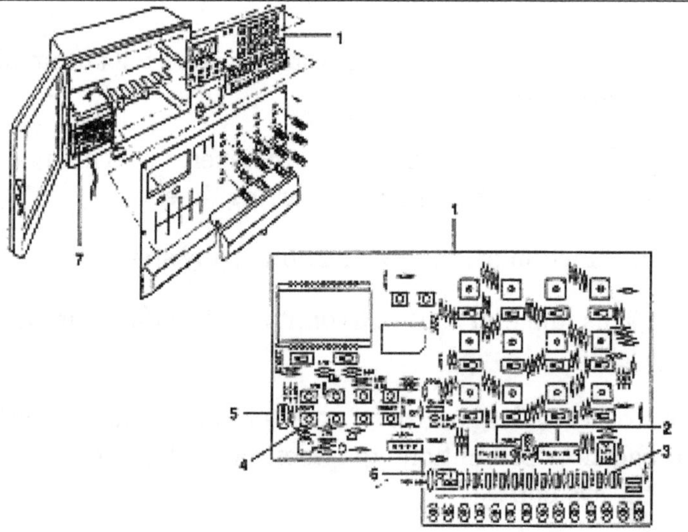

Ref.	Description	HP-6	HP-8	HP-12
1	PWB Assembly 50Hz or 60Hz	402321	402322	402323
2	I.C. 74LS138	650071	650071	650071
3	Triac, Valve Out-put side	628395	628395	628395
4	Diode 1N4002	650077-02	650077-02	650077-02
5	Capacitor, 100MF	650013-45	650013-45	650013-45
6	MOV 68v,500A	650016	650016	650016
7	Transformer 60 Hz	402320	402320	402320
7	Transformer 50 Hz	402400	402430	402430

Electronic controllers rely on solid state and Integrated circuits to provide the clock/timer, memory and control functions. These types of systems are more sensitive to powerline quality than electromechanical controllers and may be affected by spikes, surges and brownouts. Particularly spikes and surges are common in rural areas in Florida where lightning tends to be frequent and intense. These types of systems may require electrical suppression devices in order to operate reliably. Because of the inherent flexibility of electronic devices, these controllers tend to be very flexible and provide a large number of features at a relatively low cost. Figure shows the components of a commercially available electronic controller (irrigation timer).

Computer-based Irrigation Control Systems

A computer-based control system consists of a combination of hardware and software that acts as a supervisor with the purpose of managing irrigation and other related practices such as fertigation and maintenance. This is done by the use of a closed control loop. A closed control loop consists of: 1) Monitoring the state variables, 2) comparing the state variables with their desired or target state, 3) deciding what actions are necessary to change the state of the system, and 4) carrying out the necessary actions. Performing these functions requires a combination of hardware and software that must be implemented for each specific application.

Hardware Components

Figure shows the basic components of a closed loop control system, each of the hardware elements is described below.

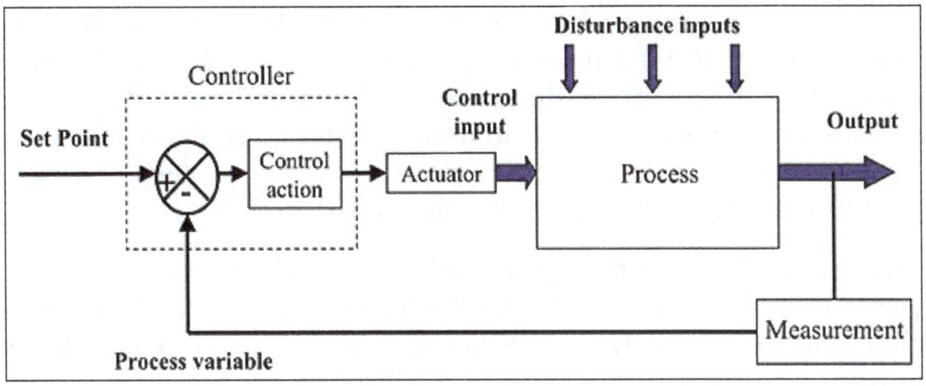

Sensors

A sensor is a device placed in the system that produces an electrical signal directly related to the parameter that is to be measured. In general, there are two types of sensors, continuous and discrete:

1. Continuous Sensors:

Continuous sensors produce a continuous electrical signal, such as a voltage, current, conductivity, capacitance, or any other measurable electrical property. For example, sensors of different kinds can be used to measure temperature, such as thermistors and thermocouples. A thermocouple will produce a voltage difference that increases as the temperature increases.

Continuous sensors are used where values taken by a state variable are required and an on/off state is not sufficient, for example, to measure pressure drop across a sand filter.

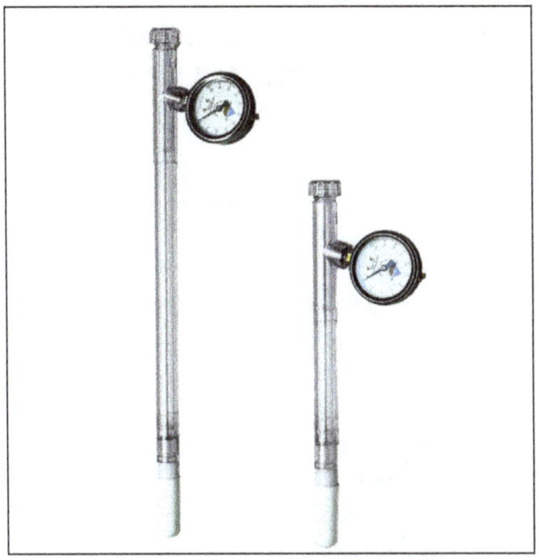

2. Discrete Sensors:

Discrete sensors are basically switches, mechanical or electronic, that indicate whether an on or off condition exists. Discrete sensors are useful for indicating thresholds, such as the opening and closure of devices (vents, doors, alarms, valves, etc). They can also be used to determine if a threshold of an important state variable has been reached. Some examples of discrete sensors are a float switch to detect if the level in a storage tank is below a minimum desirable level, a switching tensiometer to detect if soil moisture is above a desired threshold, and a thermostat to indicate if a certain temperature has been reached. When combined with time, pulses from switches can be used to measure rates. For example, tdescrete sensors can be used to measure the volume of fuel, water or chemical solution passing through a totalizing flow meter with a magnetically activated switch, or the speed of a rotating flywheel.

Sensors are an extremely important component of the control loop because they provide the basic data that drive an automatic control system. Understanding the operating principle of a sensor is very important. Sensors many times do not react directly to the variable being measured. For example, when a mercury thermometer is used to measure temperature, temperature is not being measured, rather, a change in volume due to a change in temperature is measured. Because there is a unique relationship between the volume and the temperature the instrument can be directly calibrated to provide temperature readings. The ideal sensor responds only to the "sensed" variable, without responding to any other change in the environment. It is important to understand that sensors always have a degree of inaccuracy associated with them and they may be affected by other parameters besides the "sensed" variable. The classical example is that of soil moisture measurement using electrical conductivity probes. The electrical signal produced by this sensor is closely related to soil moisture, but is greatly affected by temperature and dissolved salts (fertilizers, etc.) in the soil. Another important factor related to the sensor is its time response. A sensor must deliver a signal that reflects the state of the system within the frame of time required by the application. Using the soil moisture measurement example, the sensor must be able to "keep up" with the changes in soil moisture that are caused by evapotranspiration. Thus, proper selection of the sensors and understanding the principle of operation is critical to the success of a control system. Some of the variables that are often measured in computer based control systems are the following: 1) Flow rate, 2) pressure, 3) soil-moisture, 4) air temperature, 5) wind speed, 6) solar radiation, 7) relative humidity, 8) total salts in irrigation water, and 9) pH of irrigation water.

A/D Interface

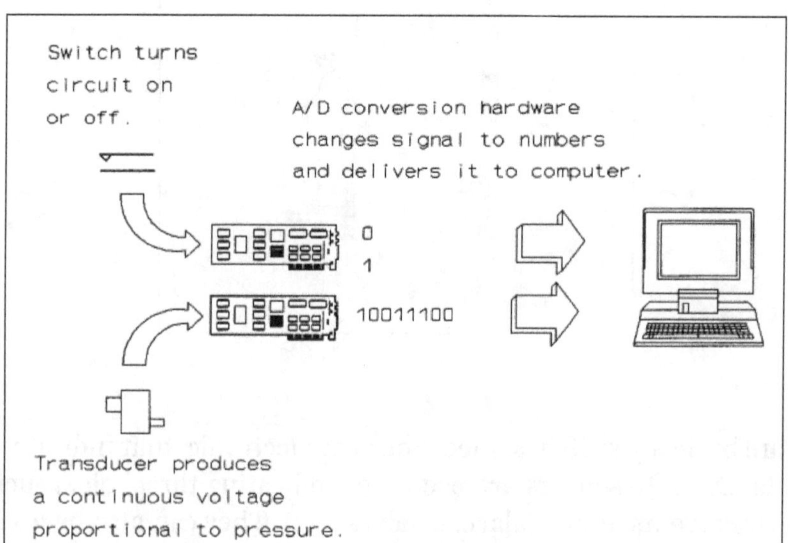

Since computer systems work internally with numbers (digits), the electrical signals resulting from the sensors must be converted to digital data. This is done through specialized hardware referred to as the Analog-to-Digital (A/D) interface. Discrete signals resulting from switch closures and threshold measurements are converted to 0 and 1. Continuous electrical (analog) signals produced by the sensors signals are converted to a number related to the level of the sensed variable. The accuracy of the conversion is affected by the resolution of the conversion equipment. In general, the higher the resolution the better the accuracy. For, example if a pressure sensor produces a voltage signal ranging from 0 to 5 volts for a range of pressure of 10 atmospheres, an 8 bit resolution A/D board will be able to detect a change in voltage of about 5/255 volts which will results in measurable increments of 10/255 atmospheres. If the resolution of the A/D board was 12 bit, the board would be able to detect a change in voltage of about 5/4095 volts or a measurable increment of 10/4095 atmospheres.

Computer System

The A/D conversion hardware is directly connected to the computer system. Given the current state of technology, the computer system may be a PC (personal computer), a minicomputer, or a specially designed machine that is solely dedicated to the control task. The type of machine depends on the type of application, and is greatly affected by factors such as environment characteristics, complexity of the controlled system, and the speed with which conversions need to take place (controlling a high speed extruder requires much more speed than a golf-course irrigation system). Many agricultural applications can be economically carried out using personal computers (PC), as is evident by the increasing number of system integrators and equipment manufactures that are marketing PC-based control systems. Also, many manufacturers of control equipment have designed and manufactured specialized computer control systems.

Control Interface

Using control software, decisions may be made to modify the controlled system. The actual changes are achieved by having devices within the system that will affect the controlled variables. These devices are controlled through actuators that respond to signals from the control interface. The devices may be of the nearly continuous or discrete types. For example, the extension of a robot arm of a citrus harvesting robot requires the use of a continuous signal from the computer, while a fan or a valve requires only an on/off (discrete) signal from the computer. In general, any device that can be powered electrically can be computer controlled.

Software Components

The software is used to implement procedures as they apply to the controlled system. These procedures are usually very elaborate, but in a well-engineered piece of software, they are transparent to the user. Because the user is more concerned with ease of use and performance of the system, good quality software has an interface that allows easy definition of the characteristics of the system to be controlled and simplifies the assignment of hardware resources. Performance is measured by how well the computer control system maintains the desired state.

Computer-based Controller Topologies

Centralized Computer

The simplest form in which a computer control system can be arranged is to use a single computer system that included the necessary support hardware and software to support data acquisition and control.

Satellite Systems

A centralized computer can be linked to other devices that have specialized purposes. One such type of device can be a datalogging system (a computer in itself) used to collect weather data. Also, in large complex systems, a central computer can be used to download instruction to intelligent controllers.

Communications between the central control computer can be implemented in a variety of ways:

- Serial or parallel communication links.
- Telephone link using serial communications.
- Carrier wave using powerline modulation.
- Communications bus.

Telephone and serial links can be hardwired or wireless. Figure shows the components of a typical high-end commercially available computer based control system.

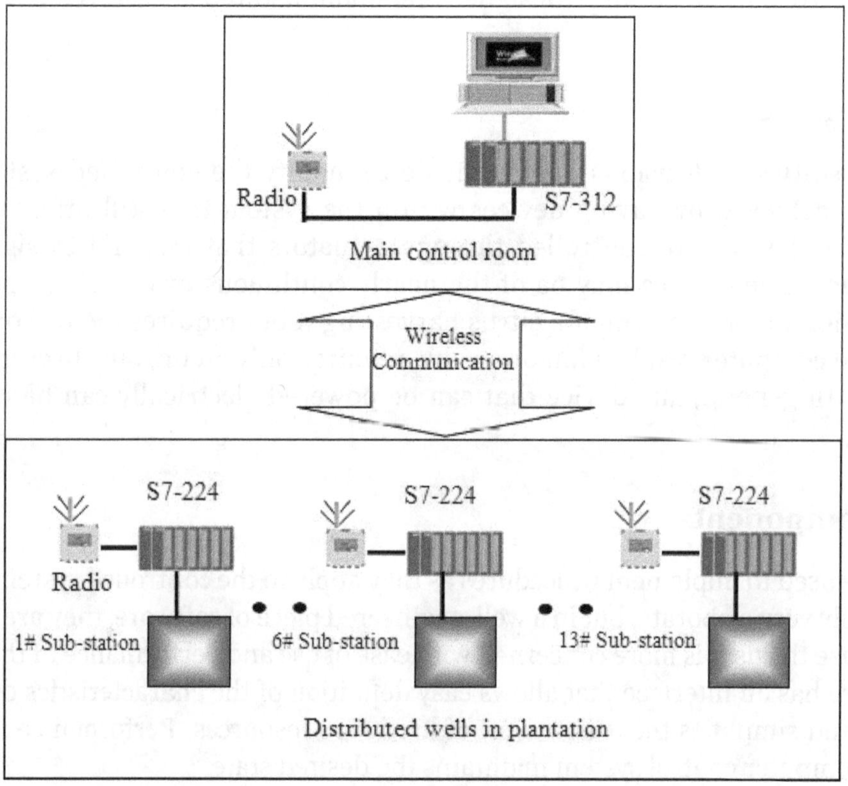

IRRIGATION SYSTEM HEADS

Irrigation heads, also known as sprinkler heads, are an integral part of many residential and commercial irrigation systems designed to distribute water to the landscape. While sprinkler heads come in myriad sizes, shapes, brands, and styles, there are four basic types of sprinkler heads:

- Pop-up spray and rotor heads.
- Impact rotors.
- Gear-driven rotors.
- Large turf rotors.

Probably the most widely used sprinkler head, pop-ups are typically used for residential and small commercial applications. Pop-up spray heads are designed to supply a continuous stream of water – nozzles are used to distribute water in a variety of patterns to fit the contours of the landscape. Typically, fixed spray heads are stationary and have no moving parts other than the "pop-up" stem in the center of the sprinkler head making them inexpensive and simple to operate.

Pop-up Spray and Rotor Heads

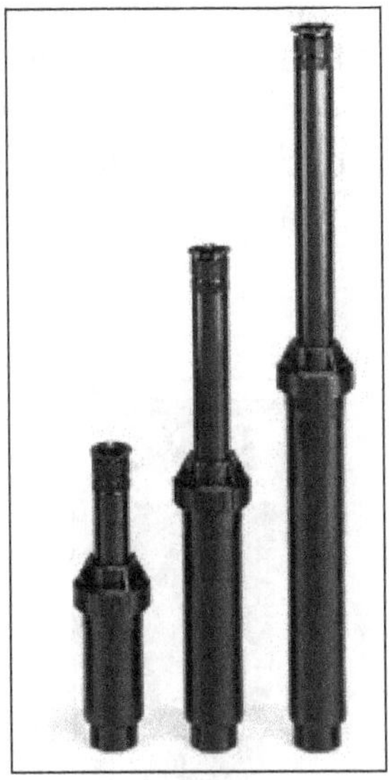

The body of pop-up heads range in height from 2 inch to 20 inches (5.1 cm to 50.8 cm). Two-inch pop-ups are usually used in situations where hard soil makes digging difficult. Four-inch pop-ups are commonly used in turf areas; the current standard for mowing height in lawn areas is typically three inches – a four-inch pop-up provides sufficient clearance for a growing lawn and the fact that over time heads tend to "sink" as soil and thatch build-up around the sprinkler head. Six to twelve-inch (15.2 cm to 30.5 cm) pop-ups can be used to irrigate groundcover, garden areas and shrub borders; heads can be installed on risers as the landscape grows and greater clearance is needed.

Pop-up spray heads are designed to be installed on a zone with plant material of similar watering needs. Each zone is operated by an individual valve that is activated by a solenoid. Pop-up spray heads are designed to cover relatively small areas with spray radius between 3 feet to 15 feet (.9 m to 4.6 m) with an operating pressure between 15 and 30 psi. The precipitation rate of fixed spray heads varies from 1.0 to 2.5 inches per hour (2.54 cm to 6.35 cm) and is dependent on the system pressure, spacing of the spray heads, manufacturer specifications, and nozzle size.

Manufacturers are continually improving pop-ups: features such as matched precipitation rates, adjustable and low trajectory nozzles, nozzles with square spacing and strips, contribute to improved efficiency of the irrigation system. Several manufacturers have even developed nozzles that convert a pop-up sprayhead to a rotor capable of covering distances ranging from 8 feet to 30 feet (2.4 m to 9.1 m). The advantage of these nozzles is that they reduce the precipitation rate, which improves infiltration and reduces run-off.

Impact Rotors

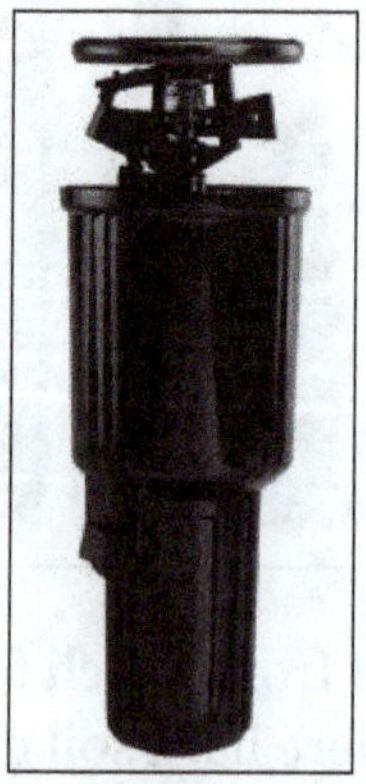

Impact rotors can be recognized by the telltale tooka, tooka, tic, tic sound made by the swinging arm striking the body of the rotor. Impact rotors provide single or multiple streams of water to the landscape and distribute water in an arc pattern typically ranging from 40° to 360°. They are designed to cover larger areas than pop-up sprayheads; the spray radius for most rotors is 20 feet to 150 feet (6.1 m to 45.7 m) with a precipitation rate between 0.1 to 1.5 inches per hour (.25 cm to 3.81 cm).

There has been very little change in the design of impact rotors over the years; its uncomplicated design makes it suitable for irrigation with well water or where other types of heads might become fouled. The radius and arc are easily adjusted, however they often require regular maintenance and the spring mechanism is often too noisy for residential sites. Because impact rotors are often made of bronze or brass they can last for many years although initially they may cost twice that of plastic, gear-driven rotors.

Gear-driven Rotors

Older impact heads are being replaced by gear-driven rotors in many situations making them the most widely used sprinkler types for medium- to large-scale areas. Their low cost, quiet operation, and versatility provide a few of the advantages that gear-driven rotors have over impact rotors. They typically require less maintenance because the enclosed body design prevents clogging of the drive mechanism from dirt and other debris.

Gear-driven rotors are best suited to small commercial sites or large residential areas. They work better than pop-up sprayheads in areas with slopes or clay because their lower precipitation rate increases infiltration into the soil.

Some manufacturers provide a nozzle tree with each of their rotors making it easy to select the appropriate size nozzle for the desired radius and precipitation rate. Typically, gear-driven rotors have a radius that ranges from 18 to 55 feet (5.5 m to 16.8 m) and an arc rotation from 40 to 360 degrees. Operating pressures range from 25 to 75 psi and precipitation rates range from 0.2 to 0.8 inches per hour (.51 cm to 2 cm) depending on the pressure, nozzle size, and the layout of the zone.

Large turf rotors.

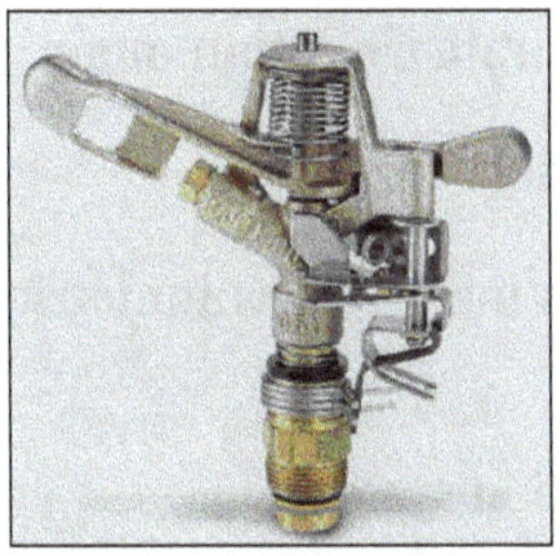

Golf course, parks, and some commercial properties require large turf rotors. These rotors require operating pressures from 50 to 100 psi. Their large size allows them to cover a radius up to 100 ft.

(30.5 m) with flows as high as 80 gallons per minute (302.7 LPM). Older model golf course heads were made of brass; some of these heads are still operating in the field 40 years after they were originally installed. Because of the high cost of brass heads the current trend is towards using plastic gear-driven rotors.

Knowing the conditions and requirements of the site to be irrigated is an important step in finding the right sprinkler head for the job. Soil type, water pressure, water source and weather conditions are just a few of the many factors that influence the type of sprinkler head needed to deliver water to the landscape. Homeowners, landscapers, and irrigation professionals benefit from a highly competitive irrigation market. Continued improvements to materials increase the durability of sprinkler heads while decreasing their costs. Advances in nozzle design can decrease overspray and runoff and reduce susceptibility to losses from wind. Although there are only four basic types of sprinkler heads, ranging from small pop-up sprayheads to large, electronically-activated, golf course rotors, they are sufficient to meet irrigation requirements of everything from a small residential landscape to a PGA golf course.

DEDICATED IRRIGATION METER

A dedicated irrigation meter is a water meter that exclusively meters water used for outdoor watering and irrigation. For many years utilities installed these meters for parks, medians, and playing fields, but more recently it has become common practice to install a separate meter for indoor and outdoor uses at many large sites with a significant irrigation demand. Even for residential properties, there is a movement to use dedicated irrigation meters.

From the utility perspective, a separate meter just for irrigation provides the following benefits:

- Accurate independent measurement of outdoor (and consequently indoor) use.
- Better ability to manage peak demands driving by irrigation.
- Ability to offer alternative pricing structures for domestic and irrigation water use.
- Improved price signaling for outdoor water use.
- Additional drought management tool.

From the customer perspective, a separate irrigation meter provides the following benefits:

- Accurate measurement of outdoor water use.
- Reduced wastewater bills as in most case volumetric wastewater charges only apply to indoor usage.
- Separate shutoff for the irrigation system.

The only real negative associated with dedicated irrigation meters is the additional cost required for purchase of the meter and installation.

In general dedicated irrigation meters are a great benefit both to water utilities and to their customers and should be encouraged.

DISC FILTER

A disc filter is a type of water filter used primarily in irrigation, similar to a screen filter, except that the filter cartridge is made of a number of plastic discs stacked on top of each other like a pile of poker chips. Each disc is covered with small grooves or bumps. The discs (or rings) each have a hole in the middle, forming a hollow cylinder in the middle of the stack. The water passes through the small passages in between and the impurities are trapped behind.

The filtration quality is based on the quantity and size of particles that the filtering element is able to retain. Higher quality filtration simply means cleaner water. This depends on the geometry of the channels, including the size, length, angle, and number of generated intersection points. The discs are typically color coded to denote the level of filtration. Filtration quality is usually measured in microns, based on the smallest size particle filtered. The typical range is from 25 microns for the finest level of filtration to 400 microns for the coarsest. Sometimes the filtration quality is given as the equivalent mesh size of a comparable screen filter. Typical mesh sizes range from 40 to 600. When using mesh sizes, 40 is the coarsest and 600 is the finest or highest level of filtration.

Disc filters range in size from small units with a 3/4" inlet and outlet used for landscape drip irrigation systems to very large banks of multiple filters manifolded together used for filtering large volumes of water for agricultural and industrial applications.

Some disc filters, especially the smaller ones, must be taken apart and cleaned by hand. Many of the larger ones can be backflushed in such a way that the discs are able to separate and spin during the cleaning cycle. In some cases, a booster pump may be required for backflushing. Disc filters can be used for many types of contaminants, including fine sand and organic matter. However, when used to filter organic matter, they will clog more quickly than a media filter and will have to be cleaned more often. One advantage that the disc filter has over the media filter is that it can backflush more quickly with less flush water. Disc filters used in agricultural irrigation are covered by the ISO 9912-2 standard.

IRRIGATION SPRINKLER

An impact sprinkler head in action.

An irrigation sprinkler (also known as a water sprinkler or simply a sprinkler) is a device used to irrigate agricultural crops, lawns, landscapes, golf courses, and other areas. They are also used for cooling and for the control of airborne dust. Sprinkler irrigation is the method of applying water in a controlled manner in way similar to rainfall. The water is distributed through a network that may consist of pumps, valves, pipes, and sprinklers.

Irrigation sprinklers can be used for residential, industrial, and agricultural usage. It is useful on uneven land where sufficient water is not available. It is useful for sandy soil. The perpendicular pipes, having rotating nozzles on top, are joined to the main pipeline at regular intervals of time. When water is allowed to flow through the main pipe under pressure with the help of pump it, escapes from the rotating nozzles. It gets sprinkled on the crop.

Types

Industrial

Higher pressure sprinklers that themselves move in a circle are driven by a ball drive, gear drive, or impact mechanism (impact sprinklers). These can be designed to rotate in a full or partial circle.

Rainguns are similar to impact sprinkler, except that they generally operate at very high pressures of 40 to 130 lbf/in^2 (275 to 900 kPa) and flows of 50 to 1200 US gal/min (3 to 76 L/s), usually with nozzle diameters in the range of 0.5 to 1.9 inches (10 to 50 mm). In addition to irrigation, guns are used for industrial applications such as dust suppression and logging.

Many irrigation sprinklers are buried in the ground along with their supporting plumbing, although above ground and moving sprinklers are also common. Most irrigation sprinklers operate through electric and hydraulic technology and are grouped together in zones that can be collectively turned on and off by actuating a solenoid-controlled valve.

Residential

An oscillating sprinkler is commonly used to water residential lawns, and is moved as needed.

Home lawn sprinklers vary widely in their size, cost, and complexity. They include impact sprinklers, oscillating sprinklers, drip sprinklers, underground sprinkler systems, and portable sprinklers. Permanently installed systems may often operate on timers or other automated processes. They are occasionally installed with retractable heads for aesthetic and practical reasons, reducing damage during lawn mowing. These types of systems usually can be programmed to automatically start on a set time and day each week.

Small portable sprinklers can be temporarily placed on lawns if additional watering is needed or if no permanent system is in place. These are often attached to an outdoor water faucet and are placed for a short period of time. Other systems may be professionally installed permanently in the ground and are attached permanently to a home's plumbing system.

An antique sprinkler developed by Nomad called a 'set-and-forget tractor sprinkler' was used in Australia in the 1950s. Water pressure ensured that the sprinkler slowly moved across a lawn.

Agricultural Science

Rotator style pivot applicator sprinkler.

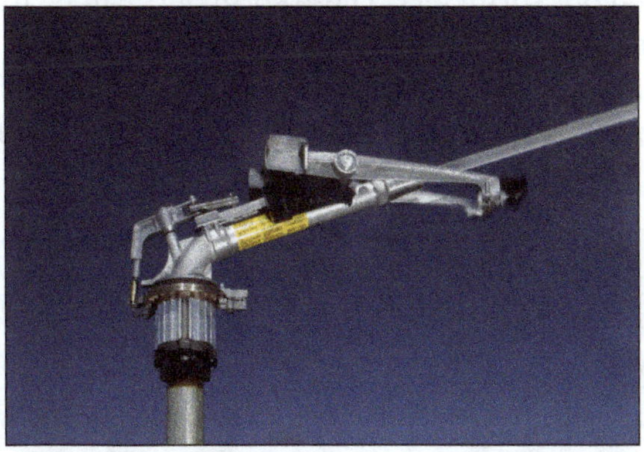

End gun style pivot applicator sprinkler.

The first use of sprinklers by farmers was some form of home and golf course type sprinklers. These ad hoc systems, while doing the job of the buried pipes and fixed sprinkler heads, interfered with cultivation and were expensive to maintain.

Center-pivot irrigation was invented in 1940 by farmer Frank Zybach, who lived in Strasburg, Colorado. In the 1950s a firm based in Portland, Oregon *Stout-Wyss Irrigation System*, developed a rolling pipe type irrigation system for farms that has become the most popular type for farmers irrigating large fields. With this system large wheels attached to the large pipes with sprinkler heads move slowly across the field.

Irrigation on a cotton farm.

Underground Sprinklers

Underground sprinklers function through means of basic electronic and hydraulic technology. This valve and all of the sprinklers that will be activated by this valve are known as a zone. Upon activation, the solenoid, which sits on top of the valve is magnetized lifting a small stainless steel plunger in its center. By doing this, the activated (or raised) plunger allows air to escape from the top of a rubber diaphragm located in the center of the valve. Water that has been charged and waiting on the bottom of this same diaphragm now has the higher pressure and lifts the diaphragm. This pressurized water is then allowed to escape down stream of the valve through a series of pipes, usually made of PVC (higher pressure commercial systems) or polyethylene pipe (for typically lower pressure residential systems). At the end of these pipes and flush to ground level (typically) are pre measured and spaced out sprinklers. These sprinklers can be fixed spray heads that have a set pattern and generally spray between 1.5–2m (7–15 ft.), full rotating sprinklers that can spray a broken stream of water from 6–12m (20–40 ft.), or small drip emitters that release a slow, steady drip of water on more delicate plants such as flowers and shrubs. use of indigenous materials also recommended.

An underground sprinkler system about to be installed.

Irrigation sprinkler watering crops.

Health Risks

In 2017, it was reported that use of common garden hoses in combination with spray nozzles may generate aerosols containing droplets smaller than 10 μm, which can be inhaled by nearby people. Water stagnating in a hose between uses, especially when warmed by the sun, can host the growth and interaction of Legionella and free-living amoebae (FLA) as biofilms on the inner surface of the hose. Clinical cases of Legionnaires' disease or Pontiac fever have been found to be associated with inhalation of garden hose aerosols containing Legionella bacteria. The report provides measured microbial densities resulting from controlled hose conditions in order to quantify the human health risks. The densities of Legionella spp. identified in two types of hoses were found to be similar to those reported during legionellosis outbreaks from other causes. It is proposed that the risk could be mitigated by draining hoses after use.

Irrigation sprinkler.

An oscillating sprinkler watering a lawn.

IMPACT SPRINKLER

An Impact Sprinkler (sometimes called an impulse sprinkler) is a type of irrigation sprinkler in which the sprinkler head, driven in a circular motion by the force of the outgoing water, pivots on a bearing on top of its threaded attachment nut. Invented in 1933 by Orton Englehart, it quickly found widespread use. Though it has in many situations been replaced by gear-driven "rotor heads", many varieties of impact sprinkler remain in use.

Development

The original horizontal action impact drive sprinkler was invented by a Glendora, California citrus grower, Orton Englehart in 1933 and patented in 1935. Orton later sold it to Clem and Mary LaFetra who manufactured and marketed it under the brand name Rain Bird.

Design and Operation

The sprinkler head pivots on a bearing on top of its threaded attachment nut. The head is driven in a circular motion by the force of the outgoing water, and at least one arm extends from the head. The sprinkler arm is repeatedly pushed back into the water stream by a spring. The arm's striking the water stream scatters the stream and re-orients the flow slightly, enabling a uniform watering area around the sprinkler.

Some full-circle impact heads feature a second, short range, opposing "spreader" nozzle, which fills the close range watering coverage role provided by the rapid "return cycle" on part-circle heads.

By adjusting the position of the limiting collars, water flow can be directed from a full-circle pattern to one of less coverage. The limiting collars are fully adjustable by twisting with fingers to achieve the area of coverage desired.

The uninterrupted flow path of impact heads makes them less vulnerable to damage and clogging by dirt and sand in the water. Thus, they are suitable for systems fed by well water. One defining feature of impact heads is they almost always have male pipe threads, as opposed to the female threads found on virtually all other sprinkler types.

The sprinkler head was originally manufactured from metal. Since the 1970s, they have also been manufactured from thermoplastics for improved corrosion resistance.

Usage

The impact sprinkler's long spray radius and uniform water distribution re-creates the effect of natural rainfall, and the device quickly replacedthe centuries-old technique of furrow irrigation, in which trenches were dug between rows of crops and flooded.

Variations

An underground pop-up version of the impact sprinkler was introduced as a way to avoid the problem of having to carefully and time consumingly mow around overground sprinkler heads. When not in operation, these sprinklers disappear out of sight below turf level; however, great care should always be shown when mowing to avoid damage to any sprinkler heads which have failed to recede back into the turf.

Although largely replaced by gear-driven "rotor heads" by the 1990s, impact sprinklers still have many advantages, including uniformity of coverage, sand and grit resistance, and operation at lower water pressures.

SPRINKLER SYSTEM TIMER

A sprinkler system timer is an electrical device that is used to set an irrigation sprinkler system to come on automatically at a certain time. Irrigation timers first appeared in the early 1960s to control large-radius lawn sprinklers, which at the time usually contained their own electrically operated valve (most golf-course sprinklers still use this type of actuation). These timers were large and cumbersome with numerous mechanical parts and were usually relegated to agricultural and commercial applications. Compact irrigation timers did not become commonplace until the 1970s when Lawn Genie introduced a mechanical timer which measured only ten by six inches and was four inches deep. This controller proved popular for many years, but was hard to reprogram and it did not operate valves in immediate succession unless each valve was set to run for an hour. Rain Bird later introduced the RC-7A to their Rain Clox line, which featured an "at a glance" electromechanical programming interface that proved very easy to operate, and offered the ability to omit stations from the program sequence without creating time gaps. This timer which became standard issue in many tract homes during the 1980s and proved to be remarkably durable in its construction, with many still operating today.

Irrigation control systems almost always use 24-volt alternating current transmitted over two wires, one of which is "common" and connected to all the valves. Other, less-common systems involve fluid-filled hydraulic tubes to open or close the valves.

Many companies followed with similar designs, such as Rainmaster, Griswold Controls, Toro, and Irritrol.

In the late 1980s, the irrigation company Hydro-Rain introduced the first "hybrid" controller design, called the HR-6100, which combined electronic programming with a visual programming interface involving a single selector dial. This overtook the electromechanical timers as the most common design, and today nearly all timers sold are hybrid designs.

The 1990s saw the introduction of computer-controlled "central control" systems, pioneered by Rain Bird for use on golf courses. This system was called "MaxiCom" and worked through a set of "cluster control units", each of which in turn synchronized a number of "satellite" controllers.

References

- Soil-moisture-sensor-intro.aspx: allianceforwaterefficiency.org, Retrieved 7 February, 2019
- Robert N. Carrow; Ronny R. Duncan; Michael T. Huck (17 December 2008). Turfgrass and Landscape Irrigation Water Quality: Assessment and Management. CRC Press. P. 275. ISBN 978-1-4200-8194-7. Retrieved 7 August 2017
- Irrigation-Controller-Introduction.aspx: allianceforwaterefficiency.org, Retrieved 9 March, 2019
- "Landscape Irrigation Equipment Part 1: Sprinklers & Spray Heads — HGIC @ Clemson University". Hgic.clemson.edu. Retrieved 2008-12-28
- Irrigation-System-Heads-Introduction.aspx: allianceforwaterefficiency.org, Retrieved 10 April, 2019
- Landers, John David (2001). Glendora California. Charleston, SC: Arcadia Publishing. Pp. 62–64. ISBN 978-0-7385-0826-9
- Dedicated-Irrigation-Meter-Introduction.aspx: allianceforwaterefficiency.org, Retrieved 11 June, 2019

6

Irrigation and the Environment

The environmental impacts of irrigation include changes in the quantity and quality of soil and water, and instability in the atmosphere. It also affects the natural and social conditions of river basins. The topics elaborated in this chapter will help in gaining a better perspective about these environmental impacts of irrigation.

IRRIGATION AND CLIMATE RESILIENCE

Many natural systems are now being affected by climate change. In order to lower the adverse impacts, the international community has adopted multiple measures to deal with climate change. Adaptation measures may soften the adverse impact of climate change, but with the intensifying climate change, adaptation measures alone can hardly remove all these impacts. Mitigation measures can reduce long-term adaptation cost to a certain extent. Therefore, to deal with climate change, it is necessary to adopt mitigation and adaption measures at the same time.

Climate change has increased the possibility and intensity of drought. So the appropriate water management would be able to increase the resilience of agricultural production to climate change. Due to the constraint of water resources, the supply of agricultural irrigation water cannot be expanded blindly.

Studies have shown that enhanced water use efficiency holds the key to tackling water scarcity and food security issues. Water saving irrigation (WSI) can mitigate the negative impact of climate change on water resources available to agriculture and overcome the constraint of water scarcity by reducing water consumption and increasing water productivity.

Studies have shown that WSI can reduce water consumption and increase grain production. Agricultural water-saving is necessary prerequisites for comprehensively redressing the worsening water shortage problems. Sprinkler irrigation is a major contributor to the significantly increased yield and water use efficiency of winter wheat than border irrigation. Large-scale water-saving crop production systems need to be established in the near future in order to feed growing population. In order to mitigate climate warming, it is necessary to minimize greenhouse gas emissions resulting from adaptation.

WSI measures help in coping with climate change, with a view to provide technical support to appropriate actions to address climate change, reduce its adverse impact on agricultural production, and ensure food security. WSI can reduce soil salinization and conserve soil to sustain land productivity and environmental benefits. So WSI can be a positive measure in coping with climate change when it is rightly deployed.

IRRIGATION WITH WASTEWATER

At the farm level, the following basic conditions should be met to make irrigated farming a success:

- The required amount of water should be applied.
- The water should be of acceptable quality.
- Water application should be properly scheduled.
- Appropriate irrigation methods should be used.
- Salt accumulation in the root zone should be prevented by means of leaching.
- The rise of water table should be controlled by means of appropriate drainage.
- Plant nutrients should be managed in an optimal way.

The above requirements are equally applicable when the source of irrigation water is treated wastewater. Nutrients in municipal wastewater and treated effluents are a particular advantage of these sources over conventional irrigation water sources and supplemental fertilizers are sometimes not necessary. However, additional environmental and health requirements must be taken into account when treated wastewater is the source of irrigation water.

Amount of Water to be Applied

It is well known that more than 99 percent of the water absorbed by plants is lost by transpiration and evaporation from the plant surface. Thus, for all practical purposes, the water requirement of crops is equal to the evapotranspiration requirement, ET_c. Crop evapotranspiration is mainly determined by climatic factors and hence can be estimated with reasonable accuracy using meteorological data. A computer program, called CROPWAT, is available in FAO to determine the water requirements of crops from climatic data. It should be kept in mind that the actual amount of irrigation water to be applied will have to be adjusted for effective rainfall, leaching requirement, application losses and other factors.

Quality of Water to be Applied

The guidelines presented are indicative in nature and will have to be adjusted depending on the local climate, soil conditions and other factors. In addition, farm practices, such as the type of crop to be grown, irrigation method, and agronomic practices, will determine to a great extent the quality suitability of irrigation water.

Table: Water Requirements, Sensitivity To Water Supply And Water Utilization Efficiency Of Some Selected Crops.

Crop	Water requirements (mm/ growing period)	Sensitivity to water supply (ky)	Water utilization efficiency for harvested yield, Ey, kg/m³ (% moisture)
Alfalfa	800-1600	low to medium-high (0.7-1.1)	1.5-2.0 hay (10-15%)
Banana	1200-2200	high (1.2-1.35)	plant crop: 2.5-4 ratoon: 3.5-6 fruit (70%)
Bean	300-500	medium-high (1.15)	lush: 1.5-2.0 (80-90%) dry: 0.3-0.6 (10%)
Cabbage	380-500	medium-low (0.95)	12-20 head (90-95%)
Citrus	900-1200	low to medium-high (0.8-1.1)	2-5 fruit (85%, lime: 70%)
Cotton	700-1300	medium-low (0.85)	0.4-0.6 seed cotton (10%)
Groundnut	500-700	low (0.7)	0.6-0.8 unshelled dry nut (15%)
Maize	500-800	high (1.25)	0.8-1.6 grain (10-13%)
Potato	500-700	medium-high (1.1)	4-7 fresh tuber (70-75%)
Rice	350-700	high	0.7-1.1 paddy (15-20%)
Safflower	600-1200	low (0.8)	0.2-0.5 seed (8-10%)
Sorghum	450-650	medium-low (0.9)	0.6-1.0 grain (12-15%)
Wheat	450-650	medium high (spring: 1.15; winter: 1.0)	0.8-1.0 grain (12-15%)

Scheduling of Irrigation

To obtain maximum yields, water should be applied to crops before the soil moisture potential reaches a level at which the evapotranspiration rate is likely to be reduced below its potential. The relationship of actual and maximum yields to actual and potential evapotranspiration is illustrated in the following equation:

$$\left(1-\frac{Y_a}{Y_m}\right)=ky\left(1-\frac{ET_a}{ET_m}\right)$$

where:

Y_a = Actual harvested yield

Y_m = Maximum harvested yield

ky = Yield response factor

ET_a = Actual evapotranspiration

ET_m = Maximum evapotranspiration

Several methods are available to determine optimum irrigation scheduling. The factors that determine irrigation scheduling are: available water holding capacity of the soils, depth of root zone, evapotranspiration rate, amount of water to be applied per irrigation, irrigation method and drainage conditions.

Irrigation Methods

Many different methods are used by farmers to irrigate crops. They range from watering individual plants from a can of water to highly automated irrigation by a centre pivot system. However, from the point of wetting the soil, these methods can be grouped under five headings, namely:

- Flood irrigation: Water is applied over the entire field to infiltrate into the soil (e.g. wild flooding, contour flooding, borders, basins, etc).
- Furrow irrigation: Water is applied between ridges (e.g. level and graded furrows, contour furrows, corrugations, etc). Water reaches the ridge, where the plant roots are concentrated, by capillary action.
- Sprinkler irrigation: Water is applied in the form of a spray and reaches the soil very much like rain (e.g. portable and solid set sprinklers, travelling sprinklers, spray guns, centre-pivot systems, etc). The rate of application is adjusted so that it does not create ponding of water on the surface.
- Sub-irrigation: Water is applied beneath the root zone in such a manner that it wets the root zone by capillary rise (e.g. subsurface irrigation canals, buried pipes, etc). Deep surface canals or buried pipes are used for this purpose.
- Localized irrigation: Water is applied around each plant or a group of plants so as to wet locally and the root zone only (e.g. drip irrigation, bubblers, micro-sprinklers, etc). The application rate is adjusted to meet evapotranspiration needs so that percolation losses are minimized.

Table: Basic features of Some selected irrigation systems.

Irrigation method	Topography	Crops	Remarks
Widely spaced borders	Land slopes capable of being graded to less than 1 % slope and preferably 0.2%	Alfalfa and other deep rooted close-growing crops and orchards	The most desirable surface method for irrigating close-growing crops where topographical conditions are favourable. Even grade in the direction of irrigation is required on flat land and is desirable but not essential on slopes of more than 0.5%. Grade changes should be slight and reverse grades must be avoided. Cross slops is permissible when confined to differences in elevation between border strips of 6-9 cm. Water application efficiency 45-60%.
Graded contour furrows	Variable land slopes of 2-25 % but preferable less	Row crops and fruit	Especially adapted to row crops on steep land, though hazardous due to possible erosion from heavy rainfall. Unsuitable for rodent-infested fields or soils that crack excessively. Actual grade in the direction of irrigation 0.5-1.5%. No grading required beyond filling gullies and removal of abrupt ridges. Water application efficiency 50-65%.
Rectangular checks (levees)	Land slopes capable of being graded so single or multiple tree basins will be levelled within 6 cm	Orchard	Especially adapted to soils that have either a relatively high or low water intake rate. May require considerable grading. Water application efficiency 40-60%.

Sub-irrigation	Smooth-flat	Shallow rooted crops such as potatoes or grass	Requires a water table, very permeable subsoil conditions and precise levelling. Very few areas adapted to this method. Water application efficiency 50-70%.
Sprinkler	Undulating 1->35% slope	All crops	High operation and maintenance costs. Good for rough or very sandy lands in areas of high production and good markets. Good method where power costs are low. May be the only practical method in areas of steep or rough topography. Good for high rainfall areas where only a small supplementary water supply is needed. Water application efficiency 60-70 %.
Localized (drip, trickle, etc.)	Any topographic condition suitable for row crop farming	Row crops or fruit	Perforated pipe on the soil surface drips water at base of individual vegetable plants or around fruit trees. Has been successfully used in Israel with saline irrigation water. Still in development stage. Water application efficiency 75-85 %.

Leaching

Under irrigated agriculture, a certain amount of excess irrigation water is required to percolate through the root zone so as to remove the salts which have accumulated as a result of evapotranspiration from the original irrigation water. This process of displacing the salts from the root zone is called leaching and that portion of the irrigation water which mobilizes the excess of salts is called the leaching fraction, LF.

$$\text{Leaching Fraction (LF)} = \frac{\text{depth of water below the root zone}}{\text{depth of water applied at the surface}}$$

Salinity control by effective leaching of the root zone becomes more important as irrigation water becomes more saline.

Drainage

Drainage is defined as the removal of excess water from the soil surface and below so as to permit optimum growth of plants. Removal of excess surface water is termed surface drainage while the removal of excess water from beneath the soil surface is termed sub-surface drainage. The importance of drainage for successful irrigated agriculture has been well demonstrated. It is particularly important in semi-arid and arid areas to prevent secondary salinization. In these areas, the water table will rise with irrigation when the natural internal drainage of the soil is not adequate. When the water table is within a few metres of the soil surface, capillary rise of saline groundwater will transport salts to the soil surface. At the surface, water evaporates, leaving the salts behind. If this process is not arrested, salt accumulation will continue, resulting in salinization of the soil. In such cases, sub-surface drainage can control the rise of the water table and hence prevent salinization.

Strategies for Managing Treated Wastewater on the Farm

Success in using treated wastewater for crop production will largely depend on adopting appropriate strategies aimed at optimizing crop yields and quality, maintaining soil productivity and safeguarding the environment. Several alternatives are available and a combination of these alternatives will offer an optimum solution for a given set of conditions. The user should have prior information on effluent supply and its quality, to ensure the formulation and adoption of an appropriate on-farm management strategy.

Basically, the components of an on-farm strategy in using treated wastewater will consist of a combination of:

- Crop selection,
- Selection of irrigation method,
- Adoption of appropriate management practices.

Furthermore, when the farmer has additional sources of water supply, such as a limited amount of normal irrigation water, he will then have an option to use both the effluent and the conventional source of water in two ways, namely:

- By blending conventional water with treated effluent,
- Using the two sources in rotation.

Table: Information required on effluent supply and quality.

Information	Decision on irrigation management
Effluent supply	
The total amount of effluent that would be made available during the crop growing season.	Total area that could be irrigated.
Effluent available throughout the year.	Storage facility during non crop growing period either at the farm or near wastewater treatment plant, and possible use for aquaculture.
The rate of delivery of effluent either as m^3 per day or litres per second.	Area that could be irrigated at any given time, layout of fields and facilities and system of irrigation.
Type of delivery: continuous or intermittent, or on demand.	Layout of fields and facilities, irrigation system, and irrigation scheduling.
Mode of supply: supply at farm gate or effluent available in a storage reservoir to be pumped by the farmer.	The need to install pumps and pipes to transport effluent and irrigation system.
Effluent quality	
Total salt concentration and electrical conductivity of the effluent.	Selection of crops, irrigation method, leaching and other management practices.
Concentrations of cations, such as Ca^{++}, Mg^{++} and Na^{+}.	To assess sodium hazard and undertake appropriate measures.
Concentration of toxic ions, such as heavy metals, Boron and Cl^{-}.	To assess toxicities that are likely to be caused by these elements and take appropriate measures.
Concentration of trace elements (particularly those which are suspected of being phyto-toxic).	To assess trace toxicities and take appropriate measures.
Concentration of nutrients, particularly nitrate-N.	To adjust fertilizer levels, avoid over-fertilization and select crop.
Level of suspended sediments.	To select appropriate irrigation system and measures to prevent clogging problems.
Levels of intestinal nematodes and faecal coliforms.	To select appropriate crops and irrigation systems.

Crop Selection

To Overcome Salinity Hazards

Not all plants respond to salinity in a similar manner; some crops can produce acceptable yields at much higher soil salinity than others. This is because some crops are better able to make the needed osmotic adjustments, enabling them to extract more water from a saline soil. The ability of a crop to adjust to salinity is extremely useful. In areas where a build-up of soil salinity cannot be controlled at an acceptable concentration for the crop being grown, an alternative crop can be selected that is both more tolerant of the expected soil salinity and able to produce economic yields. There is an 8-10 fold range in the salt tolerance of agricultural crops. This wide range in tolerance allows for greater use of moderately saline water, much of which was previously thought to be unusable. It also greatly expands the acceptable range of water salinity (ECw) considered suitable for irrigation.

The relative salt tolerance of most agricultural crops is known well enough to give general salt tolerance guidelines. The following general conclusions can be drawn from these data:

- Full yield potential should be achievable with nearly all crops when using a water with salinity less than 0.7 dS/m.
- When using irrigation water of slight to moderate salinity (i.e. 0.7-3.0 dS/m), full yield potential is still possible but care must be taken to achieve the required leaching fraction in order to maintain soil salinity within the tolerance of the crops. Treated sewage effluent will normally fall within this group.
- For higher salinity water (more than 3.0 dS/m) and sensitive crops, increasing leaching to satisfy a leaching requirement greater than 0.25 to 0.30 might not be practicable because of the excessive amount of water required. In such a case, consideration must be given to changing to a more tolerant crop that will require less leaching, to control salts within crop tolerance levels. As water salinity (ECw) increases within the slight to moderate range, production of more sensitive crops may be restricted due to the inability to achieve the high leaching fraction needed, especially when grown on heavier, more clayey soil types.

If the salinity of the applied water exceeds 3.0 dS/m, the water might still be usable but its use may need to be restricted to more permeable soils and more salt-tolerant crops, where high leaching fractions are more easily achieved. This is being practised on a large scale in the Arabian Gulf States, where drip irrigation systems are widely used.

If the exact cropping patterns or rotations are not known for a new area, the leaching requirement must be based on the least tolerant of the crops adapted to the area. In those instances, where soil salinity cannot be maintained within acceptable limits of preferred sensitive crops, changing to more tolerant crops will raise the area's production potential. If there is any doubt about the effect of wastewater salinity on crop production, a pilot study should be undertaken to demonstrate the feasibility of irrigation and the outlook for economic success.

To Overcome Toxicity Hazards

A toxicity problem is different from a salinity problem in that it occurs within the plant itself and is not caused by water shortage. Toxicity normally results when certain ions are taken up by plants with the soil water and accumulate in the leaves during water transpiration to such an extent that the plant is damaged. The degree of damage depends upon time, concentration of toxic material, crop sensitivity and crop water use and, if damage is severe enough, crop yield is reduced. Common toxic ions in irrigation water are chloride, sodium, and boron, all of which will be contained in sewage. Damage can be caused by each individually or in combination. Not all crops are equally sensitive to these toxic ions. Some guidance on the sensitivity of crops to sodium, chloride and boron. However, toxicity symptoms can appear in almost any crop if concentrations of toxic materials are sufficiently high. Toxicity often accompanies or complicates a salinity or infiltration problem, although it may appear even when salinity is not a problem.

The toxic ions of sodium and chloride can also be absorbed directly into the plant through the leaves when moistened during sprinkler irrigation. This typically occurs during periods of high temperature and low humidity. Leaf absorption speeds up the rate of accumulation of a toxic ion and may be a primary source of the toxicity.

In addition to sodium, chloride and boron, many trace elements are toxic to plants at low concentrations. Fortunately, most irrigation supplies and sewage effluents contain very low concentrations of these trace elements and are generally not a problem.

However, urban wastewater may contain heavy metals at concentrations which will give rise to elevated levels in the soil and cause undesirable accumulations in plant tissue and crop growth reductions. Heavy metals are readily fixed and accumulate in soils with repeated irrigation by such wastewaters and may either render them non-productive or the product unusable. Surveys of wastewater use have shown that more than 85 % of the applied heavy metals are likely to accumulate in the soil, most at the surface. The levels at which heavy metals accumulation in the soil is likely to have a deleterious effect on crops. Any wastewater use project should include monitoring of soil and plants for toxic materials.

To Prevent Health Hazards

From the point of view of human consumption and potential health hazards, crops and cultivated plants may be classified into the following groups:

Table: Relative Tolerance Of Selected Crops To Exchangeable Sodium.

Sensitive	Semi-tolerant	Tolerant
Avocado	Carrot	Alfalfa
(Persea americana)	(Daucus carota)	(Medicago sativa)
Deciduous Fruits	Clover, Ladino	Barley
Nuts	(Trifolium repens)	(Hordeum vulgare)
Bean, green	Dallisgrass	Beet, garden
(Phaseolus vulgaris)	(Paspalum dilatatum)	(Beta vulgaris)
Cotton (at germination)	Fescue, tall	Beet, sugar
(Gossypium hirsutum)	(Festuca arundinacea)	(Beta vulgaris)
Maize	Lettuce	Bermuda grass
(Zea mays)	(Lactuca sativa)	(Cynodon dactylon)

Peas	Bajara	Cotton
(Pisum sativum)	(Pennisetum typhoides)	(Gossypium hirsutum)
Grapefruit	Sugarcane	Paragrass
(Citrus paradisi)	(Saccharum officinarum)	(Brachiaria mutica)
Orange	Berseem	Rhodes grass
(Citrus sinensis)	(Trifolium alexandrinum)	(Chloris gayana)
Peach	Benji	Wheatgrass, crested
(Prunus persica)	(Mililotus parviflora)	(Agropyron cristatum)
Tangerine	Raya	Wheatgrass, fairway
(Citrus reticulata)	(Brassica juncea)	(agropyron cristatum)
Mung	Oat	Wheatgrass, tall
(Phaseolus aurus)	(Avena sativa)	(Agropyron elongatum)
Mash	Onion	Karnal grass
(Phaseolus mungo)	(Allium cepa)	(Diplachna fusca)

- Food crops:
  - Those eaten uncooked.
  - Those eaten after cooking.
- Forage and feed crops:
  - Direct access by animals.
  - Those fed to animals after harvesting.

Table: Chloride tolerance of some fruit crop cultivars and rootstocks.

Crop	Rootstock or Cultivar	Maximum permissible Cl- without leaf injury	
		Root zone (Cl_e) (me/l)	Irrigation water (Cl_w) (me/l)
	Rootstocks		
Avocado (Persea americana)	West Indian	7.5	5.0
	Guatemalan	6.0	4.0
	Mexican	5.0	3.3
Citrus (Citrus spp.)	Sunki Mandarin	25.0	16.6
	Grapefruit		
	Cleopatra mandarin		
	Rangpur lime		
	Sampson tangelo	15.0	10.0
	Rough lemon		
	Sour orange		
	Ponkan mandarin		

	Citrumelo 4475	10.0	6.7
	Trifoliate orange		
	Cuban shaddock		
	Calamondin		
	Sweet orange		
	Savage citrange		
	Rusk citrange		
	Troyer citrange		
Grape(Vitis spp.)	Salt Creek, 1613-3	40.0	27.0
	Dog Ridge	30.0	20.0
Stone Fruits (Prunus spp.)	Marianna	25.0	17.0
	Lovell, Shalil	10.0	6.7
	Yunnan	7.5	5.0
	Cultivars		
Berries (Rubus spp.)	Boysenberry	10.0	6.7
	Olallie clackberry	10.0	6.7
	Indian SUmmer	5.0	3.3
	Raspberry		
Grape(Vitis spp.)	Thompson seedless	20.0	13.3
	Perlette	20.0	13.3
	Cardinal	10.0	6.7
	Black Rose	10.0	6.7
Strawberry (Fragaria spp.)	Lassen	7.5	5.0
	Shasta	5.0	3.3

1. For some crops, the concentration given may exceed the overall salinity tolerance of that crop and cause some reduction in yield in addition to that caused by chloride ion toxicities.
2. Values given are for the maximum concentration in the irrigation water. The values were derived from saturation extract data (EC_e) assuming a 15-20 percent leaching fraction and $EC_d = 1.5\ EC_w$.
3. The maximum permissible values apply only to surface irrigated crops. Sprinkler irrigation may cause excessive leaf bum at values far below these.

Maximum concentrations tolerated in soil water without yield or vegetative growth reductions. Boron tolerances vary depending upon climate, soil conditions and crop varieties. Maximum concentrations in the irrigation water are approximately equal to these values or slightly less.

- Landscaping plants:
  - Unprotected areas with public access.
  - Semi-protected areas.

- Afforestation plants:
    - Commercial (fruit, timber, fuel and charcoal).
    - Environmental protection (including sand stabilization).

In terms of health hazards, treated effluent with a high microbiological quality is necessary for the irrigation of certain crops, especially vegetable crops eaten raw, but a lower quality is acceptable for other selected crops, where there is no exposure to the public.

Example: Categorization of Crops in Relation to Exposed Group and Health Control Measures.

Category A:

- Protection required for consumers, agricultural workers, and the general public.
- Includes crops likely to be eaten uncooked, spray-irrigated fruits and grass (sports fields, public parks and lawns).

Category B:

- Protection required for agricultural workers only.
- Includes cereal crops, industrial crops (such as cotton and sisal), food crops for canning, fodder crops, pasture and trees.
- In certain circumstances some vegetable crops might be considered as belonging to Category B if they are not eaten raw (potatoes, for instance) or if they grow well above ground (for example, chillies), in such cases it is necessary to ensure that the crop is not contaminated by sprinkler irrigation or by falling on to the ground, and that contamination of kitchens by such crops, before cooking, does not give rise to a health risk.

Selection of Irrigation Methods

Under normal conditions, the type of irrigation method selected will depend on water supply conditions, climate, soil, crops to be grown, cost of irrigation method and the ability of the farmer to manage the system. However, when using wastewater as the source of irrigation other factors, such as contamination of plants and harvested product, farm workers, and the environment, and salinity and toxicity hazards, will need to be considered. There is considerable scope for reducing the undesirable effects of wastewater use in irrigation through selection of appropriate irrigation methods.

The choice of irrigation method in using wastewater is governed by the following technical factors:

- The choice of crops.
- The wetting of foliage, fruits and aerial parts.
- The distribution of water, salts and contaminants in the soil.
- The ease with which high soil water potential could be maintained.

- The efficiency of application.
- The potential to contaminate farm workers and the environment.

Table presents an analysis of these factors in relation to four widely practised irrigation methods, namely border, furrow, sprinkler and drip irrigation.

Table: Evaluation of common irrigation methods in relation to the use of treated wastewater.

Parameters of evaluation	Furrow irrigation	Border irrigation	Sprinkler irrigation	Drip irrigation
Foliar wetting and consequent leaf damage resulting in poor yield	No foliar injury as the crop is planted on the ridge	Some bottom leaves may be affected but the damage is not so serious as to reduce yield	Severe leaf damage can occur resulting in significant yield loss	No foliar injury occurs under this method of irrigation
Salt accumulation in the root zone with repeated applications	Salts tend to accumulate in the ridge which could harm the crop	Salts move vertically downwards and are not likely to accumulate in the root zone	Salt movement is downwards and root zone is not likely to accumulate salts	Salt movement is radial along the direction of water movement. A salt wedge is formed between drip points
Ability to maintain high soil water potential	Plants may be subject to stress between irrigations	Plants may be subject . to water stress between irrigations	Not possible to maintain high soil water potential throughout the growing season	Possible to maintain high soil water potential throughout the growing season and minimize the effect of salinity
Suitability to handle brackish wastewater without significant yield loss	Fair to medium. With good management and drainage acceptable yields are possible	Fair to medium. Good irrigation and drainage practices can produce acceptable levels of yield	Poor to fair. Most crops suffer from leaf damage and yield is low	Excellent to good. Almost all crops can be grown with very little reduction in yield

A border (and basin or any flood irrigation) system involves complete coverage of the soil surface with treated effluent and is normally not an efficient method of irrigation. This system will also contaminate vegetable crops growing near the ground and root crops and will expose farm workers to the effluent more than any other method. Thus, from both the health and water conservation points of view, border irrigation with wastewater is not satisfactory.

Furrow irrigation, on the other hand, does not wet the entire soil surface. This method can reduce crop contamination, since plants are grown on the ridges, but complete health protection cannot be guaranteed. Contamination of farm workers is potentially medium to high, depending on automation. If the effluent is transported through pipes and delivered into individual furrows by means of gated pipes, risk to irrigation workers will be minimum.

The efficiency of surface irrigation methods in general, borders, basins, and furrows, is not greatly affected by water quality, although the health risk inherent in these systems is most certainly of concern. Some problems might arise if the effluent contains large quantities of suspended solids and these settle out and restrict flow in transporting channels, gates, pipes and appurtenances. The use of primary treated sewage will overcome many of such problems. To avoid surface ponding of stagnant effluent, land levelling should be carried out carefully and appropriate land gradients should be provided.

Sprinkler, or spray, irrigation methods are generally more efficient in terms of water use since greater uniformity of application can be achieved. However, these overhead irrigation methods may contaminate ground crops, fruit trees and farm workers. In addition, pathogens contained in aerosolized effluent may be transported downwind and create a health risk to nearby residents. Generally, mechanized or automated systems have relatively high capital costs and low labour costs compared with manually-moved sprinkler systems. Rough land levelling is necessary for sprinkler systems, to prevent excessive head losses and achieve uniformity of wetting. Sprinkler systems are more affected by water quality than surface irrigation systems, primarily as a result of the clogging of orifices in sprinkler heads, potential leaf burns and phytotoxicity when water is saline and contains excessive toxic elements, and sediment accumulation in pipes, valves and distribution systems. Secondary wastewater treatment has generally been found to produce an effluent suitable for distribution through sprinklers, provided that the effluent is not too saline. Further precautionary measures, such as treatment with granular filters or micro-strainers and enlargement of nozzle orifice diameters to not less than 5 mm, are often adopted.

Localized irrigation, particularly when the soil surface is covered with plastic sheeting or other mulch, uses effluent more efficiently, can often produce higher crop yields and certainly provides the greatest degree of health protection for farm workers and consumers. Trickle and drip irrigation systems are expensive, however, and require a high quality of effluent to prevent clogging of the emitters through which water is slowly released into the soil. Table presents water quality requirements to prevent clogging in localized irrigation systems. Solids in the effluent or biological growth at the emitters will create problems but gravel filtration of secondary treated effluent and regular flushing of lines have been found to be effective in preventing such problems in Cyprus. Bubbler irrigation, a technique developed for the localized irrigation of tree crops avoids the need for small emitter orifices but careful setting is required for its successful application.

Table: Water Quality and Clogging Potential in Drip Irrigation Systems.

Potential Problem	Units	Degree of Restriction on Use		
	None	Slight to Moderate	Severe	
Physical				
Suspended Solids	mg/l	< 50	50- 100	> 100
Chemical				
pH		< 7.0	7.0 - 8.0	> 8.0
Dissolved Solids	mg/l	< 500	500-2000	> 2000
Manganese	mg/l	< 0.1	0.1 - 1.5	> 1.5
Iron	mg/l	< 0.1	0.1 - 1.5	> 1.5
Hydrogen Sulphide	mg/l	< 0.5	0.5 - 2.0	> 2.0
Biological	maximum			
Bacterial populations	number/ml	< 10000	10 000 - 50 000	> 50000

When compared with other systems, the main advantages of trickle irrigation seem to be:

- Increased crop growth and yield achieved by optimizing the water, nutrients and air regimes in the root zone.

- High irrigation efficiency: No canopy interception, wind drift or conveyance losses and minimal drainage losses.
- Minimal contact between farm workers and effluent.
- Low energy requirements: The trickle system requires a water pressure of only 100-300 k Pa (1-3 bar).
- Low labour requirements: The trickle system can easily be automated, even to allow combined irrigation and fertilization (sometimes terms fertigation).

Apart from the high capital costs of trickle irrigation systems, another limiting factor in their use is that they are only suited to the irrigation of row crops. Relocation of subsurface systems can be prohibitively expensive.

Clearly, the decision on irrigation system selection will be mainly a financial one but it is to be hoped that the health risks associated with the different methods will be taken into account. The method of effluent application is one of the health control measures possible, along with crop selection, wastewater treatment and human exposure control. Each measure will interact with the others and thus a decision on irrigation system selection will have an influence on wastewater treatment requirements, human exposure control and crop selection (for example, row crops are dictated by trickle irrigation). At the same time the irrigation techniques feasible will depend on crop selection and the choice of irrigation system might be limited if wastewater treatment has already been decided before effluent use is considered.

Field Management Practices in Wastewater Irrigation

Management of water, soil, crop and operational procedures, including precautions to protect farm workers, play an important role in the successful use of sewage effluent for irrigation.

Water Management

Most treated wastewaters are not very saline, salinity levels usually ranging between 500 and 200 mg/l (ECw = 0.7 to 3.0 dS/m). However, there may be instances where the salinity concentration exceeds the 2000 mg/l level. In any case, appropriate water management practices will have to be followed to prevent salinization, irrespective of whether the salt content in the wastewater is high or low. It is interesting to note that even the application of a non-saline wastewater, such as one containing 200 to 500 mg/l, when applied at a rate of 20,000 m^3 per hectare, a fairly typical irrigation rate, will add between 2 and 5 tonnes of salt annually to the soil. If this is not flushed out of the root zone by leaching and removed from the soil by effective drainage, salinity problems can build up rapidly. Leaching and drainage are thus two important water management practices to avoid salinization of soils.

Leaching

The concept of leaching has already been discussed. The question that arises is how much water should be used for leaching, i.e. what is the leaching requirement? To estimate the leaching requirement, both the salinity of the irrigation water (EC_w) and the crop tolerance to soil salinity (EC_e) must be known. The necessary leaching requirement (LR) can be estimated from figure for general crop rotations reported by Ayers and Westcot. A more exact estimate of the leaching requirement for a particular crop can be obtained using the following equation:

$$LR = \frac{EC_w}{5(EC_e - EC_w)}$$

where:

LR = Minimum leaching requirement needed to control salts within the tolerance (EC_e) of the crop with ordinary surface methods of irrigation

EC_w = Salinity of the applied irrigation water in dS/m.

EC_e = Average soil salinity tolerated by the crop as measured on a soil saturation extract. It is recommended that the EC_e value that can be expected to result in at least a 90% or greater yield be used in the calculation.

Figure was developed using EC_e values for the 90% yield potential. For water in the moderate to high salinity range (>1.5 dS/m), it might be better to use the EC_e value for maximum yield potential (100%) since salinity control is critical in obtaining good yields.

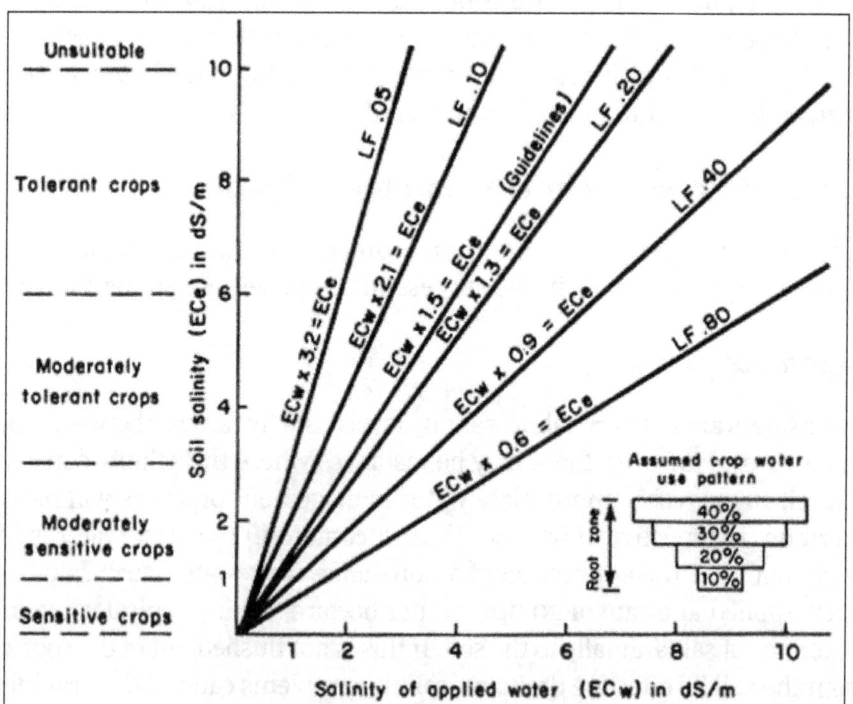

Relationship between applied water salinity and soil water salinity at different leaching.

Where water is scarce and expensive, leaching practices should be designed to maximize crop production per unit volume of water applied, to meet both the consumptive use and leaching requirements. Depending on the salinity status, leaching can be carried out at each irrigation, each alternative irrigation or less frequently, such as seasonally or at even longer intervals, as necessary to keep the salinity in the soil below the threshold above which yield might be affected to an unacceptable level. With good quality irrigation water, the irrigation application level will almost always apply sufficient extra water to accomplish leaching. With high salinity irrigation water, meeting the leaching requirement is difficult and requires large amounts of water. Rainfall must be considered in estimating the leaching requirement and in choosing the leaching method.

The following practices are suggested for increasing the efficiency of leaching and reducing the amount of water needed:

1. Leach during cool seasons instead of during warm periods, to increase the efficiency and ease of leaching, since the total annual crop water demand (ET, mm/year) losses are lower.

2. Use more salt: Tolerant crops which require a lower leaching requirement (LR) and thus have a lower water demand.

3. Use tillage to slow overland water flow and reduce the number of surface cracks which bypass flow through large pores and decrease leaching efficiency.

4. Use sprinkler irrigation at an application rate below the soil infiltration rate as this favours unsaturated flow, which is significantly more efficient for leaching than saturated flow. More irrigation time but less water is required than for continuous ponding.

5. Use alternate ponding and drying instead of continuous ponding as this is more efficient for leaching and uses less water, although the time required to leach is greater. This may have drawbacks in areas having a high water table, which allows secondary salinization between pondings.

6. Where possible, schedule leachings at periods of low crop water use or postpone teachings until after the cropping season.

7. Avoid fallow periods, particularly during hot summers, when rapid secondary soil salinization from high water tables can occur.

8. If infiltration rates are low, consider pre-planting irrigations or off-season leaching to avoid excessive water applications during the crop season.

9. Use one irrigation before the start of the rainy season if total rainfall is normally expected to be insufficient for a complete leaching. Rainfall is often the most efficient leaching method because it provides high quality water at relatively low rates of application.

Drainage

Salinity problems in many irrigation projects in arid and semi-arid areas are associated with the presence of a shallow water table. The role of drainage in this context is to lower the water table to a desirable level, at which it does not contribute to the transport of salts to the root zone and the soil surface by capillarity. What is important is to maintain a downward movement of water through soils. van Schilfgaard reported that drainage criteria are frequently expressed in terms of critical water table depths; although this is a useful concept, prevention of salinization depends on the establishment, averaged over a period of time, of a downward flux of water. Another important element of the total drainage system is its ability to transport the desired amount of drained water out of the irrigation scheme and dispose of it safely. Such disposal can pose a serious problem, particularly when the source of irrigation water is treated wastewater, depending on the composition of the drainage effluent.

Timing of Irrigation

The timing of irrigation, including irrigation frequency, pre-planting irrigation and irrigation prior to a winter rainy season, can reduce the salinity hazard and avoid water stress between irrigations. Some of these practices are readily applicable to wastewater irrigation.

In terms of meeting the water needs of crops, increasing the frequency of irrigation will be desirable as it eliminates water stress between irrigations. However, from the point of view of overall water management, this may not always produce the desired results. For example, with border, basin and other flood irrigation methods, frequent irrigations may result in an unacceptable increase in the quantity of water applied, decrease in water use efficiency and larger amounts of water to be drained. However, with sprinklers and localized irrigation methods, frequent applications with smaller amounts may not result in decrease in water use efficiency and, indeed, could help to overcome the salinity problem associated with saline irrigation water.

Pre-planting irrigation is practised in many irrigation schemes for two reasons, namely: (i) to leach salts from the soil surface which may have accumulated during the previous cropping period and to provide a salt-free environment to germinating seeds (it should be noted that for most crops, the seed germination and seedling stages are most sensitive to salinity); and (ii) to provide adequate moisture to germinating seeds and young seedlings. A common practice among growers of lettuce, tomatoes and other vegetable crops is to pre-irrigate the field before planting, since irrigation soon after planting could create local water stagnation and wet spots that are not desirable. Treated wastewater is a good source for pre-irrigation as it is normally not saline and the health hazards are practically nil.

Blending of Wastewater with other Water Supplies

One of the options that may be available to farmers is the blending of treated sewage with conventional sources of water, canal water or ground water, if multiple sources are available. It is possible that a farmer may have saline ground water and, if he has non-saline treated wastewater, could blend the two sources to obtain a blended water of acceptable salinity level. Further, by blending, the microbial quality of the resulting mixture could be superior to that of the unblended wastewater.

Alternating Treated Wastewater with other Water Sources

Another strategy is to use the treated wastewater alternately with the canal water or groundwater, instead of blending. From the point of view of salinity control, alternate applications of the two sources will be superior to blending. However, an alternating application strategy will require duel conveyance systems and availability of the effluent dictated by the alternate schedule of application.

Land and Soil Management

Several land and soil management practices can be adopted at the field level to overcome salinity, sodicity, toxicity and health hazards that might be associated with the use of treated wastewater.

Land Development

During the early stages of on-farm land development, steps can be taken to minimize potential hazards that may result from the use of wastewater. These will have to be well planned, designed and executed since they are expensive and, often, one time operations. Their goal is to improve permanently existing land and soil conditions in order to make irrigation with wastewater easier. Typical activities include levelling of land to a given grade, establishing adequate drainage (both open and sub-surface systems), deep ploughing and leaching to reduce soil salinity.

Land Grading

Land grading is important to achieve good uniformity of application from surface irrigation methods and acceptable irrigation efficiencies in general. If the wastewater is saline, it is very important that the irrigated land is appropriately graded. Salts accumulate in the high spots which have too little water infiltration and leaching, while in the low spots water accumulates, causing waterlogging and soil crusting.

Land grading is well accepted as an important farm practice in irrigated agriculture. Several methods are available to grade land to a desired slope. The slope required will vary with the irrigation system, length of run of water flow, soil type, and the design of the field. Recently, laser techniques have been applied to level land precisely so as to obtain high irrigation efficiencies and prevent salinization.

Deep Cultivation

In certain areas, the soil is stratified, and such soils are difficult to irrigate. Layers of clay, sand or hard pan in stratified soils frequently impede or prevent free movement of water through and beyond the root zone. This will not only lead to saturation of the root zone but also to accumulation of salts in the root zone. Irrigation efficiency as well as water movement in the soil can be greatly enhanced by sub-soiling and chiselling of the land. The effects of sub-soiling and chiselling remain for about 1 to 5 years but, if long term effects are required, the land should be deep and slip ploughed. Deep or slip ploughing is costly and usually requires the growing of annual crops soon after to allow the settling of the land. Following a couple of grain crops, grading will be required to re-establish a proper grade to the land.

Crop Management and Cultural Practices

Several cultural and crop management practices that are valid under saline water use will be valid under wastewater use. These practices are aimed at preventing damage to crops caused by salt accumulation surrounding the plants and in the root zone and adjusting fertilizer and agrochemical applications to suit the quality of the wastewater and the crop.

Placement of Seed

In most crops, seed germination is more seriously affected by soil salinity than other stages of development of a crop. The effects are pronounced in furrow-irrigated crops, where the water is fairly to highly saline. This is because water moves upwards by capillarity in the ridges, carrying salts with it. When water is either absorbed by roots or evaporated, salts are deposited in the ridges. Typically, the highest salt concentration occurs in the centre of the ridge, whereas the lowest concentration of salt is found along the shoulders of the ridges. An efficient means of overcoming this problem is to ensure that the soil around the germinating seeds is sufficiently low in salinity. Appropriate planting methods, ridge shapes and irrigation management can significantly decrease damage to germinating seeds. Some specific practices include:

1. Planting on the shoulder of the ridge in the case of single row planting or on both shoulders in double row planting,
2. Using sloping beds with seeds planted on the sloping side, but above the water line,
3. Irrigating alternate rows so that the salts can be moved beyond the single seed row.

Planning for Wastewater Irrigation

Central Planning

Government policy on effluent use in agriculture will have a deciding effect on what control measures can be achieved through careful selection of site and crops to be irrigated with treated effluent. A decision to make treated effluent available to farmers for unrestricted irrigation or to irrigate public parks and urban green areas with effluent will remove the possibility of taking advantage of careful selection of sites, irrigation techniques and crops in limiting the health risks and minimizing environmental impacts. However, if a Government decides that effluent irrigation will only be applied in specific controlled areas, even if crop selection is not limited (that is, unrestricted irrigation is allowed within these areas), public access to the irrigated areas will be prevented and can be applied. Without doubt, the greatest security against health risk and adverse environmental impact will be achieved by limiting effluent use to restricted irrigation on controlled areas to which the public has no access but even imposing restrictions on effluent irrigation by farmers, if properly enforced, can achieve a degree of control.

Cobham and Johnson have suggested that the procedures involved in preparing plans for effluent irrigation schemes are similar to those used in most forms of resource planning and summarized the main physical, social and economic dimensions. They also indicated that a number of key issues or tasks were likely to have a significant effect on the ultimate success of effluent irrigation, as follows:

1. Organizational and managerial provisions made to administer the resource, to select the effluent use plan and to implement it,
2. The importance attached to public health considerations and the levels of risk taken,
3. The choice of single-use or multiple-use strategies,
4. The criteria adopted in evaluating alternative reuse proposals,
5. The level of appreciation of the scope for establishing a forest resource.

Adopting a mix of effluent use strategies is normally advantageous in respect of allowing greater flexibility, increased financial security and more efficient use of the wastewater throughout the year, whereas a single-use strategy will give rise to seasonal surpluses of effluent for unproductive disposal. Therefore, in site and crop selection the desirability of providing areas for different crops and forestry so as to utilize the effluent at maximum efficiency over the whole yearly cycle of seasons must be kept in mind.

Desirable Site Characteristics

The features which are critical in deciding the viability of a land disposal project are the location of available land and public attitudes. Land which is far distant from the sewage treatment plant will incur high costs for transporting treated effluent to site and will generally not be suitable. Hence, the availability of land for effluent irrigation should be considered when sewerage is being planned and sewage treatment plants should be strategically located in relation to suitable agricultural sites. Ideally, these sites should not be close to residential areas but even remote land might not be acceptable to the public if the social, cultural or religious attitudes are opposed to the practice of wastewater irrigation. The potential health hazards associated with effluent irrigation can make this a very sensitive issue and public concern will only be mollified by the application of strict control measures. In arid areas, the importance of agricultural use of treated effluent makes it advisable to be as systematic as possible in planning, developing and managing effluent irrigation projects and the public must be kept informed at all stages.

The ideal objective in site selection is to find a suitable area where long-term application of treated effluent will be feasible without adverse environmental or public health impacts. It might be possible in a particular instance to identify several potential sites within reasonable distance of the sewered community and the problem will be to select the most suitable area or areas, taking all relevant factors into account. The following basic information on an area under consideration will be of value, if available:

- A topographic map,
- Agricultural soils surveys,
- Aerial photographs,
- Geological maps and reports,
- Groundwater reports and well logs,
- Boring logs and soil test results,
- Other soil and peizometric data.

At this preliminary stage of investigation it should be possible to assess the potential impact of treated effluent application on any usable aquifer in the area(s) concerned. The first ranking of sites should take into account other factors, such as the cost and location of the land, its present use and availability, and social factors, in addition to soil and groundwater conditions.

The characteristics of the soil profile underlying a particular site are very important in deciding on its suitability for effluent irrigation and the methods of application to be employed. Among the soil properties important from the point of view of wastewater application and agricultural production are: physical parameters (such as texture, grading, liquid and plastic limits, etc.), permeability, water-holding capacity, pH, salinity and chemical composition. Preliminary observation of sites, which could include shallow hand-auger borings and identification of vegetation, will often allow the elimination of clearly unsatisfactory sites. After elimination of marginal sites, each site under serious consideration must be investigated by on-site borings to ascertain the soil profile, soil characteristics and location of the water table. Peizometers should be located in each borehole and these can be used for subsequent groundwater sampling. A procedure for such site assessment has been described by Hall and Thompson and, if applied, should not only allow the most suitable site among several possible to be selected but permit the impact of effluent irrigation at the chosen site to be modelled. When a site is developed, a long-term groundwater monitoring programme should be an essential feature of its management.

Crop Selection Issues

Normally, in choosing crops, a farmer is influenced by economics, climate, soil and water characteristics, management skill, labour and equipment available and tradition. The degree to which the use of treated effluent influences crop selection will depend on Government policy on effluent irrigation, the goals of the user and the effluent quality. Government policy will have the objectives of minimizing the health risk and influencing the type of productivity associated with effluent irrigation. Regulations must be realistic and achievable in the context of national and local environmental conditions and traditions. At the same time, planners of effluent irrigation schemes must attempt to achieve maximum productivity and water conservation through the choice of crops and effluent application systems.

A multiple-use strategy approach will require the evaluation of viable combinations of the cropping options possible on the land available. This will entail a considerable amount of survey and resource budgeting work, in addition to the necessary soil and water quality assessments. The annual, monthly and daily water demands of the crops, using the most appropriate irrigation techniques, have to be determined. Domestic consumption, local production and imports of the various crops must be assessed so that the economic potential of effluent irrigation of the various crop combinations can be estimated. Finally, the crop irrigation demands must be matched with the available effluent so as to achieve optimum physical and financial utilization throughout the year. This process of assessment is reviewed by Cobham and Johnson for the case of effluent use in Kuwait, where afforestation for commercial purposes was found to offer significant potential in multiple-use effluent irrigation.

ENVIRONMENTAL IMPACT OF IRRIGATION

The environmental impacts of irrigation relate to the changes in quantity and quality of soil and water as a result of irrigation and the effects on natural and social conditions in river basins and downstream of an irrigation scheme. The impacts stem from the altered hydrological conditions caused by the installation and operation of the irrigation scheme.

Direct Effects

An irrigation scheme draws water from groundwater, rivers, lakes or overland flow, and distributes it over an area. Hydrological, or direct, effects of doing this include reduction in downstream river flow, increased evaporation in the irrigated area, increased level in the water table as groundwater recharge in the area is increased and flow increased in the irrigated area. Likewise, irrigation has immediate effects on the provision of moisture to the atmosphere, inducing atmospheric instabilities and increasing downwind rainfall, or in other cases modifies the atmospheric circulation, delivering rain to different downwind areas. Increases or decreases in irrigation are a key area of concern in precipitationshed studies, that examine how significant modifications to the delivery of evaporation to the atmosphere can alter downwind rainfall.

Indirect Effects

Indirect effects are those that have consequences that take longer to develop and may also be longer-lasting. The indirect effects of irrigation include the following:

- Waterlogging
- Soil salination
- Ecological damage
- Socioeconomic impacts

The indirect effects of waterlogging and soil salination occur directly on the land being irrigated. The ecological and socioeconomic consequences take longer to happen but can be more far-reaching.

Some irrigation schemes use water wells for irrigation. As a result, the overall water level decreases. This may cause water mining, land/soil subsidence, and, along the coast, saltwater intrusion.

Irrigated land area worldwide occupies about 16% of the total agricultural area and the crop yield of irrigated land is roughly 40% of the total yield. In other words, irrigated land produces 2.5 times more product than non-irrigated land. This article will discuss some of the environmental and socioeconomic impacts of irrigation.

Adverse Impacts

Reduced River Flow

The reduced downstream river flow may cause:

- Reduced downstream flooding.
- Disappearance of ecologically and economically important wetlands or flood forests.
- Reduced availability of industrial, municipal, household, and drinking water.
- Reduced shipping routes. Water withdrawal poses a serious threat to the Ganges. In India, barrages control all of the tributaries to the Ganges and divert roughly 60 percent of river flow to irrigation.
- Reduced fishing opportunities. The Indus River in Pakistan faces scarcity due to over-extraction of water for agriculture. The Indus is inhabited by 25 amphibian species and 147 fish species of which 22 are found nowhere else in the world. It harbors the endangered Indus River dolphin, one of the world's rarest mammals. Fish populations, the main source of protein and overall life support systems for many communities, are also being threatened.

- Reduced discharge into the sea, which may have various consequences like coastal erosion (e.g. in Ghana) and salt water intrusion in delta's and estuaries (e.g. in Egypt, see Aswan dam). Current water withdrawal from the river Nile for irrigation is so high that, despite its size, in dry periods the river does not reach the sea. The Aral Sea has suffered an "environmental catastrophe" due to the interception of river water for irrigation purposes.

Increased Groundwater Recharge, Waterlogging and Soil Salinity

Looking over the shoulder of a Peruvian farmer in the Huarmey delta at waterlogged and salinised irrigated land with a poor crop stand.

This illustrates an environmental impact of upstream irrigation developments causing an increased flow of groundwater to this lower-lying area, leading to adverse conditions.

Increased groundwater recharge stems from the unavoidable deep percolation losses occurring in the irrigation scheme. The lower the irrigation efficiency, the higher the losses. Although fairly high irrigation efficiencies of 70% or more (i.e. losses of 30% or less) can occur with sophisticated techniques like sprinkler irrigation and drip irrigation, or by well managed surface irrigation, in practice the losses are commonly in the order of 40% to 60%. This may cause the following issues:

- Rising water tables.
- Increased storage of groundwater that may be used for irrigation, municipal, household and drinking water by pumping from wells.
- Waterlogging and drainage problems in villages, agricultural lands, and along roads - with mostly negative consequences. The increased level of the water table can lead to reduced agricultural production.
- Shallow water tables: A sign that the aquifer is unable to cope with the groundwater recharge stemming from the deep percolation losses.
- Where water tables are shallow, the irrigation applications are reduced. As a result, the soil is no longer leached and soil salinity problems develop.
- Stagnant water tables at the soil surface are known to increase the incidence of water-borne diseases like malaria, filariasis, yellow fever, dengue, and schistosomiasis (Bilharzia) in many areas. Health costs, appraisals of health impacts and mitigation measures are rarely part of irrigation projects, if at all.

- To mitigate the adverse effects of shallow water tables and soil salinization, some form of watertable control, soil salinity control, drainage and drainage system is needed.
- As drainage water moves through the soil profile it may dissolve nutrients (either fertilizer-based or naturally occurring) such as nitrates, leading to a buildup of those nutrients in the ground-water aquifer. High nitrate levels in drinking water can be harmful to humans, particularly infants under 6 months, where it is linked to "blue-baby syndrome".

Reduced Downstream River Water Quality

Owing to drainage of surface and groundwater in the project area, which waters may be salinized and polluted by agricultural chemicals like biocides and fertilizers, the quality of the river water below the project area can deteriorate, which makes it less fit for industrial, municipal and household use. It may lead to reduced public health. Polluted river water entering the sea may adversely affect the ecology along the sea shore.

The natural buildup of sedimentation can reduce downstream river flows due to the installation of irrigation systems. Sedimentation is an essential part of the ecosystem that requires the natural flux of the river flow. This natural cycle of sediment dispersion replenishes the nutrients in the soil, that will in turn, determine the livelihood of the plants and animals that rely on the sediments carried downstream. The benefits of heavy deposits of sedimentation can be seen in large rivers like the Nile River. The sediment from the delta has built up to form a giant aquifer during flood season, and retains water in the wetlands. The wetlands that are created and sustained due to built up sediment at the basin of the river is a habitat for numerous species of birds. However, heavy sedimentation can reduce downstream river water quality and can exacerbate floods up stream. This has been known to happen in the Sanmenxia reservoir in China. The Sanmenxia reservoir is part of a larger man-made project of hydro-electric dams called the Three Gorge Project In 1998, uncertain calculations and heavy sediment greatly affected the reservoir's ability to properly fulfill its flood-control function This also reduces the downstream river water quality. Shifting more towards mass irrigation installments in order to meet more socioeconomic demands is going against the natural balance of nature, and use water pragmatically-use it where it is found.

Affected Downstream Water Users

Water becomes scarce for nomadic pastoralist in Baluchistan due to new irrigation developments.

Downstream water users often have no legal water rights and may fall victim of the development of irrigation. Pastoralists and nomadic tribes may find their land and water resources blocked by new irrigation developments without having a legal recourse.

Flood-recession cropping may be seriously affected by the upstream interception of river water for irrigation purposes.

- In Baluchistan, Pakistan, the development of new small-scale irrigation projects depleted the water resources of nomadic tribes traveling annually between Baluchistan and Gujarat or Rajasthan, India.
- After the closure of the Kainji dam, Nigeria, 50 to 70 per cent of the downstream area of flood-recession cropping was lost.

Lake Manantali, 477 km^2, displaced 12,000 people.

Lost Land use Opportunities

Irrigation projects may reduce the fishing opportunities of the original population and the grazing opportunities for cattle. The livestock pressure on the remaining lands may increase considerably, because the ousted traditional pastoralist tribes will have to find their subsistence and existence elsewhere, overgrazing may increase, followed by serious soil erosion and the loss of natural resources.The Manatali reservoir formed by the Manantali dam in Mali intersects the migration routes of nomadic pastoralists and destroyed 43000 ha of savannah, probably leading to overgrazing and erosion elsewhere. Further, the reservoir destroyed 120 km^2 of forest. The depletion of groundwater aquifers, which is caused by the suppression of the seasonal flood cycle, is damaging the forests downstream of the dam.

Groundwater Mining with Wells and Land Subsidence

Flooding as a consequence of land subsidence.

When more groundwater is pumped from wells than replenished, storage of water in the aquifer is being mined and the use of that water is no longer sustainable. As levels fail, it becomes more difficult to extract water and pumps will struggle to maintain the design flowrate and consume more may fenergy per unit of water. Eventually it may become so difficult to extract groundwater that farmers may be forced to abandon irrigated agriculture. Some notable examples include:

- The hundreds of tube wells installed in the state of Uttar Pradesh, India, with World Bank funding have operating periods of 1.4 to 4.7 hours/day, whereas they were designed to operate 16 hours/day.
- In Baluchistan, Pakistan, the development of tubewell irrigation projects was at the expense of the traditional qanat or karez users.
- Groundwater-related subsidence of the land due to mining of groundwater occurred in the United States at a rate of 1m for each 13m that the watertable was lowered.
- Homes at Greens Bayou near Houston, Texas, where 5 to 7 feet of subsidence has occurred, were flooded during a storm in June 1989 as shown in the picture.

Simulation and Prediction

The effects of irrigation on watertable, soil salinity and salinity of drainage and groundwater, and the effects of mitigative measures can be simulated and predicted using agro-hydro-salinity models like SaltMod and SahysMod.

- In India 2.19 million have been reported to suffer from waterlogging in irrigation canal commands. Also 3.47 million ha were reported to be seriously salt affected.
- In the Indus Plains in Pakistan, more than 2 million hectares of land is waterlogged. The soil of 13.6 million hectares within the Gross Command Area was surveyed, which revealed that 3.1 million hectares (23%) was saline. 23% of this was in Sindh and 13% in the Punjab. More than 3 million ha of water-logged lands have been provided with tube-wells and drains at the cost of billions of rupees, but the reclamation objectives were only partially achieved. The Asian Development Bank (ADB) states that 38% of the irrigated area is now waterlogged and 14% of the surface is too saline for use.
- In the Nile delta of Egypt, drainage is being installed in millions of hectares to combat the water-logging resulting from the introduction of massive perennial irrigation after completion of the High Dam at Assuan.
- In Mexico, 15% of the 3 million ha of irrigable land is salinized and 10% is waterlogged.
- In Peru some 0.3 million ha of the 1.05 million ha of irrigable land suffers from degradation.
- Estimates indicate that roughly one-third of the irrigated land in the major irrigation countries is already badly affected by salinity or is expected to become so in the near future. Present estimates for Israel are 13% of the irrigated land, Australia 20%, China 15%, Iraq 50%, Egypt 30%. Irrigation-induced salinity occurs in large and small irrigation systems alike.

- FAO has estimated that by 1990 about 52 million ha of irrigated land will need to have improved drainage systems installed, much of it subsurface drainage to control salinity.

Reduced Downstream Drainage and Groundwater Quality

- The downstream drainage water quality may deteriorate owing to leaching of salts, nutrients, herbicides and pesticides with high salinity and alkalinity. There is threat of soils converting into saline or alkali soils. This may negatively affect the health of the population at the tail-end of the river basin and downstream of the irrigation scheme, as well as the ecological balance. The Aral Sea, for example, is seriously polluted by drainage water.
- The downstream quality of the groundwater may deteriorate in a similar way as the downstream drainage water and have similar consequences.

Mitigation of Adverse effects

Irrigation can have a variety negative impacts on ecology and socioeconomy, which may be mitigated in a number of ways. These include siting the irrigation project in a location which minimises negative impacts. The efficiency of existing projects can be improved and existing degraded croplands can be improved rather than establishing a new irrigation project Developing small-scale, individually owned irrigation systems as an alternative to large-scale, publicly owned and managed schemes. The use of sprinkler irrigation and micro-irrigation systems decreases the risk of waterlogging and erosion. Where practicable, using treated wastewater makes more water available to other users Maintaining flood flows downstream of the dams can ensure that an adequate area is flooded each year, supporting, amongst other objectives, fishery activities.

Delayed Environmental Impacts

It often takes time to accurately predict the impact that new irrigation schemes will have on the ecology and socioeconomy of a region. By the time these predictions are available, a considerable amount of time and resources may have already been expended in the implementation of that project. When that is the case, the project managers will often only change the project if the impact would be considerably more than they had originally expected.

Potential Benefits Outweigh the Potential Disadvantages

Frequently irrigation schemes are seen as extremely necessary for socioeconomic well-being especially in developing countries. One example of this can be demonstrated from a proposal for an irrigation scheme in Malawi. Here it was shown that the potential positive effects of the irrigation project that was being proposed "outweighed the potential negative impacts". It was stated that the impacts would mostly "be localized, minimal, short term occurring during the construction and operation phases of the Project". In order to help alleviate and prevent major environmental impacts, they would use techniques that minimize the potential negative impacts. As far as the region's socioeconomic well-being, there would be no "displacement and resettlement envisioned during the implementation of the Project activities". The original primary purposes of the irrigation project were to reduce poverty, improve food security, create local employment, increase household income and enhance the sustainability of land use.

Due to this careful planning this project was successful both in improving the socialeconomic conditions in the region and ensuring that land and water are sustainability into the future.

POLLUTION AND IRRIGATION

It is well known that agriculture is the single largest user of freshwater resources, using a global average of 70% of all surface water supplies. Except for water lost through evapotranspiration, agricultural water is recycled back to surface water and groundwater. The water that is lost through evapotranspiration leaves its share of salts in the irrigation water that returns to the other downstream water bodies. In addition, it carries with it various chemicals from fertilizers, pesticides and other additives used in crop production. Through desalinization and drainage practices to relieve waterlogging of irrigated land, the drainage water, rich in salts also poses a challenge, if not disposed of in a planned manner. At the same time there is a net loss of soil due to poor agricultural practices.

With increasing use of recycled wastewater, especially in developing countries of arid and semi-arid regionsand its likely impact to public health, food safety and environmental quality. Particularly in cases where the wastewater is a mix of domestic as well as industrial effluents, the presence of certain chemicals have impact not only the water but also the productivity of soil.

Pollution by sediment has two major dimensions – physical, and chemical.Sediment, as a physical pollutant, impacts receiving waters in the following ways. High levels ofturbiditylimit penetration of sunlight into the water column, thereby limiting or prohibiting growth of algae and rooted aquatic plants. The hypertrophic (nutrient rich) status of many shallow lakes, especially in developing countries, would give rise to immense growth of algae and rooted plants were it not for the limiting effect of light extinction due to high turbidity. High levels ofsedimentationin rivers lead to physical disruption of the hydraulic characteristics of the channel.

The role of sediment in chemical pollution is tied both to the particle size of sediment, and to the amount of particulate organic carbon associated with the sediment. Organic chemicals associated with sediment enter into the food chain in a variety of ways. Ultimately, toxic compounds bio-accumulate in fish and other top predators. In this way, pesticides that are transported off the land as part of the runoff and erosion process accumulate in top predators including man.

References

- O. A. Tuinenburg et al., the fate of evaporated water from the ganges basin, journal of geophysical research: atmospheres, volume 117, issue d1, 16 January 2012
- Res-irri-climate: icid.org, Retrieved 30 March, 2019
- Bruce sundquist, 2007. Chapter 1- irrigation overview. In: the earth's carrying capacity, some related reviews and analysis. On line: "archived copy". Archived from the original on 2012-02-17. Retrieved 2012-02-17
- Res-irri-pollution: icid.org, Retrieved 1 June, 2019
- N.t. singh, 2005. Irrigation and soil salinity in the indian subcontinent: past and present. Lehigh university press. Isbn 0-934223-78-5, isbn 978-0-934223-78-2, 404

7

Waterlogging

The saturation of soil with water is referred to as waterlogging. Various methods are used to prevent water logging such as land grading, tree planting, reduction in unnecessary percolation of rainfall, efficient drainage system, in-field water management, etc. This chapter closely examines the key concepts of waterlogging to provide an extensive understanding of the subject.

When the conditions are so created that the crop root-zone gets deprived of proper aeration due to the presence of excessive moisture or water content, the tract is said to be waterlogged. To create such conditions it is not always necessary that under groundwater table should enter the crop root-zone. Sometimes even if water table is below the root-zone depth the capillary water zone may extend in the root-zone depth and makes the air circulation impossible by filling the pores in the soil.

The waterlogging may be defined as rendering the soil unproductive and infertile due to excessive moisture and creation of anaerobic conditions. The phenomenon of waterlogging can be best understood with the help of a hydrologic equation, which states that,

Inflow = Outflow -I- Storage

Here inflow represents that amount of water which enters the subsoil in various processes. It includes seepage from the canals, infiltration of rainwater, percolation from irrigated fields and subsoil flow. Thus although it is loss or us, it represents the amount of water flowing into the soil.

The term outflow represents mainly evaporation from soil, transpiration from plants and underground drainage of the tract. The term storage represents the change in the groundwater reservoir.

Causes of Waterlogging

After studying the phenomenon of waterlogging in the light of hydrologic equation main factors which help in raising the water-table may be recognised correctly.

- Inadequate drainage of over-land run-off increases the rate of percolation and in turn helps in raising the water table.
- The water from rivers may infiltrate into the soil.
- Seepage of water from earthen canals also adds significant quantity of water to the underground reservoir continuously.

- Sometimes subsoil does not permit free flow of subsoil water which may accentuate the process of raising the water table.
- Irrigation water is used to flood the fields. If it is used in excess it may help appreciably in raising the water table. Good drainage facility is very essential.

Effects of Waterlogging

The waterlogging affects the land in various ways. The various after effects are the following:

- Creation of Anaerobic Condition in the Crop Root-Zone: When the aeration of the soil is satisfactory bacteriological activities produce the required nitrates from the nitrogenous compounds present in the soil. It helps the crop growth. Excessive moisture content creates anaerobic condition in the soil. The plant roots do not get the required nourishing food or nutrients. As a result crop growth is badly affected.
- Growth of Water Loving Wild Plants: When the soil is waterlogged water loving wild plant life grows abundantly. The growth of wild plants totally prevent the growth of useful crops.
- Impossibility of Tillage Operations: Waterlogged fields cannot be tilled properly. The reason is that the soil contains excessive moisture content and it does not give proper tilth.
- Accumulation of Harmful Salts: The upward water movement brings the toxic salts in the crop root-zone. Excess accumulation of these salts may turn the soil alkaline. It may hamper the crop growth.
- Lowering of Soil Temperature: The presence of excessive moisture content lowers the temperature of the soil. In low temperature the bacteriological activities are retarded which affects the crop growth badly.
- Reduction in Time of Maturity: Untimely maturity of the crops is the characteristic of waterlogged lands. Due to this shortening of crop period the crop yield is reduced considerably.

Detection of Waterlogging

It is clear that the waterlogging is indicated when the ground water reservoir goes on building up continuously. When the storage starts building up in the initial stages the crop growth is actually increased because more water is made available for the crop growth. But after some time the waters table rises very high and the land gets waterlogged. Finally the land is rendered unproductive and infertile.

The problem of waterlogging develops in its full form slowly. Therefore its early detection is possible by keeping a close watch over the yields and also on the variations in the groundwater level. A comparative reduction in crop yields in spite of irrigation and fertilisation and early maturity of crops indicate the symptoms of waterlogging. Also when harmful salts start appearing on the fields as white incrustation or deposit it indicates that waterlogging is likely to follow. In worst cases the water-table rises so high and close to the ground surface that the fields turn into swamps and marshes.

The best way of keeping watch over the problem of waterlogging is by observing variations in the groundwater level. It can be done by measuring the depth of water levels at regular interval in the wells dug in the area. Continuous high water levels indicate that the groundwater storage is building up which may create waterlogging in the area.

Solution to the Problem of Waterlogging

The problem of waterlogging may be attacked on two fronts. First is preventive measures, which keep the land free from waterlogging. Secondly curative measures may be adopted to reclaim the waterlogged area. But in principle both measures aim at reducing the inflow and augmenting the outflow from the underground reservoir.

Preventive Measures

Preventive measures include the following:

- Controlling the loss of water due to seepage from the canals. The seepage loss may be reduced by adopting various measures for example:
  - By lowering the FSL of the canal: Loss may be due to percolation or absorption but when FSL is lowered the loss is reduced to sufficient extent. It is course essential to see that while lowering the FSL command is not sacrificed.
  - By lining the canal section: When the canal section is made fairly watertight by providing lining the seepage loss is reduced to quite a good extent.
  - By introducing intercepting drains: They are generally constructed parallel to the canal. They give exceptionally good results for the reach where the canal runs in high embankments.
- Preventing the loss of water due to percolation from field channels and fields: The percolation loss can be removed by using water more economically. It may also be affected by keeping intensity of irrigation low. Then only small portion of the irrigable tract is flooded and consequently the percolation loss takes place only on the limited area. It keeps the water-table sufficiently low.
- Augmentation of outflow and prevention of inflow: It may be accomplished by introducing artificial open and underground drainage grid. It may also be achieved by improving the flow conditions of existing natural drainages.
- Quick disposal of rainwater: Quick removal of rainwater by surface or open drains is a very effective method of preventing the rise in water table and consequent waterlogging of the tract. It is needless to state that the rainwater removed is net reduction in inflow.

Curative Measures

Curative measures include the following:

Installation of lift irrigation systems: When a lift irrigation project in the form of a tube well irrigation system is introduced in the waterlogged area the water table gets lowered sufficiently. It is found to be very successful method of reclaiming waterlogged land. Thus a combination of a canal system and a supplementary tube well irrigation system may be considered to be most successful and efficient irrigation scheme.

Of course it is true that it will create some complications while assessing the charges for irrigation water. (The canal water being cheaper than tube well water). Implementation of drainage schemes: The waterlogged area may be reclaimed by introducing overland and underground drainage schemes.

Implementation of Drainage Schemes: The waterlogged area may be reclaimed by introducing overland and underground drainage schemes. In our country water-logging is a problem of great concern. It is estimated that total area of waterlogged land is 86.92 lakh hectares. It includes area in irrigation commands as well as other area outside the command.

State	Water-logged area in lakh	
	Total	In irrigated commands
Andhra Pradesh	3.39	2.66
Assam	4.50	N.R.
Bihar	7.07	6.20
Gujarat	4.84	1.73
Haryana	6.20	2.49
Jammu and Kashmir	0.10	0.02
Kerala	0.61	0.12
Karnataka	0.25	0.25
Madhya Pradesh	0.73	0.73
Maharashtra	1.11	0.16
Orissa	1.96	1.96
Punjab	10.90	2.00
Rajasthan	3.48	1.80
Tamil Nadu	0.18	0.16
Uttar Pradesh	19.80	4.30
West Bengal	21.80	N.R.
Total	86.92	24.57

While the areas in the irrigation command get waterlogged due to rise in water table as a direct consequence of inadequate drainage, other areas get waterlogged due to inundation, as consequence of flooding for long durations. The States mainly affected and the extent of area rendered infertile and unproductive are given in table.

Steps are being taken to reclaim the waterlogged land in the country. The steps taken to reclaim such areas include implementation of drainage schemes, provision of deep drains, excavation of new channels and improvement of existing ones, construction of sluices with marginal embankment and installation of tube wells.

The spread of conjunctive use of groundwater with that of surface water especially in Punjab, Haryana and parts of Uttar Pradesh has substantially lowered the groundwater table and helped in containing water-logging/salinity.

Summarising the most effective and efficient anti-water-logging measures are:

1. Lining of channels (main canal, branches and field channels).
2. Provision of surface drains for drainage of rainwater.
3. Implementation of tube well projects both extensive and local.

Other methods are of local importance and hence benefit derived from them is limited.

REDUCTION OF WATERLOGGING IN AGRICULTURAL LANDS

Agricultural lands become waterlogged or poorly drained mostly as a result of poor water management, either under rainfed or irrigated agriculture. By simply adopting various designs and proper land husbandry, waterlogging can be reduced or even eliminated. These measures are presented here categorized for both rainfed and irrigated systems.

Preventing Waterlogging in Irrigated Agriculture

Irrigation without proper drainage sometimes increases the total water held in a soil profile to an extent of causing poor drainage. This is controlled through reduction of excessive water inflow into the soil as follows:

Control of Canal Seepage

Canal seepage is a major source of water losses, and waterlogging in irrigated areas. It can be controlled by:

- Lining of canals with impervious material e.g. concrete, clay, polythene, so as to control excessive seepage from unlined canals is a very important preventive measure.
- Lowering of designed full supply level to the canal, with a view to cut down the seepage losses. Increasing the gradient of the canal t enable faster flows hence less ponding.
- Interception drains along the canals 15 to 30 metres away from the canal banks are effective to arrest seepage from the canals.
- If possible, convert water conveyance system from canals to piped systems.
- Regulating canal discharge: Seepage losses can be minimized if canal discharge is regulated during nonirrigational period. Overuse of canal water is curbed if canal supply is regulated together with pumping of ground water.

Unlined canal can cause waterlogging of adjacent lands.

Lined canal stops waterlogging by preventing seepage.

In-field Water Management

This is perhaps the most important aspect of avoiding the percolation of unnecessary water in irrigated lands. It entails:

- Applying only the requisite amount of irrigation water (not to exceed the field capacity), so that all the water applied is used by plants.
- Adopting more efficient irrigation methods e.g. drip irrigation.
- Where feasible, convert from surface irrigation methods to sprinkler or drip irrigation.
- Restrict irrigation in areas with high water table to receive supplemental irrigation during dry season, or only a fraction of total command area.
- Economical use of water like leveling the fields and refraining from applying more water than the soil can hold.

Control of Inflow from other Water Bodies

Flood water can overrun irrigated fields from overflowing rivers, flood plains and channel obstruction. Such flooding should be stopped from reaching the low-lying fields.

This can be achieved through:

- Construction of dykes to train the river movement.
- On- stream dams, which act like regulating dams.
- Off-stream dams.
- To conserve the catchment with run off collection and disposal systems.

Control of Ground-water Inflows

Interflow water is mainly from sloping ground rivers, high infiltration rate and a can cause increased waterlogging.

These are controlled by:

- Construction of an interceptor.
- Growing trees e.g. eucalyptus or crops that have high water consumption rates (e.g. sugarcane, arrow roots).
- Construction of spring protection works.

Control of Water from Higher Ground into Lower Fields

Excessive water inflows from higher ground can sometimes inundate irrigated lowlands through ground water recharge.

The solution would be to:

- Obtain the water for irrigation from somewhere else.
- Proper irrigation water management.
- Incorporate fish or duck farming.

Preventative Measures in Rainfed Agriculture

Rainfed agriculture usually suffers from temporary waterlogging associated with excessive rainfall, flat topography or poor land and water management. Poor drainage can be controlled using preventative measures such as:

Reducing Unnecessary Percolation of Rainfall

Excessive runoff onto flat areas cause waterlogging.

Cambered beds can be used to reduce waterlogging.

This involves efficient disposal of storm water in excess of crop water requirement. The excess water collects because of high seasonal rainfall (termed in-situ rainfall), concentrated over a short period, in an extensive flat land, or downstream channel obstruction, or soils of shallow impermeable barrier (planosols) or low infiltration rates (vertisols). The solution would be to have:

- Cambered beds.
- Leveling of flat areas to reduce depressions.
- Sub-soiling to break the hard pans and improve infiltrations.
- Excavation of farm ponds to store water for later and use/supplementary irrigation.
- Grow water loving plants e.g. sugarcane.
- Improving terrace outlets before the rains. A functional outlet is the most important part of the drainage system.

Controlling Surface Runoff

Surface runoff from large catchments or paved areas such as homesteads, urban areas and roads can be excessive to the point of causing waterlogging of receiving lowlands. The excess water can be due to high intensity storms on sloping ground, high ratio of catchment to receiving areas or poor land use and management practices in the catchment. The solution would be to:

- Construct diversion ditches/cutoff drains to intercept the runoff.
- Conservation in the catchment areas to reduce runoff accumulation.
- Water harvesting and storage in tanks, pans and other structures.
- Developing a functional runoff collection and disposal system.

Tree planting

Tree planting, particularly of species that take up a lot of water, e.g. eucalyptus tree species, can be helpful in lowering water table. The trees further ameliorate the environment and can be used for commercial purposes. Deep rooted eucalyptus trees have roots extending up to 3 m but transpire water at a high rate and thus works as a biological pump. This they can take care of unwanted drainage water in irrigation schemes and temporary waterlogging in rainfed systems. But the bio-drainage is only effective if there is adequate aeration in the root zone and the roots extend up to ground water reservoir and draw water from capillary zone.

Other Preventative Measures

- Land grading: Land grading is useful for improving surface drainage in waterlogged soils, especially in irrigated lands.
- Cropping patterns: Vegetation and crops which have high rate of evapotranspiration may be patronized to serve as natural drainage system.
- Optimum water use: Conjunctive use of surface and ground water be encouraged by digging wells and sinking shallow tube wells.
- Better irrigation methods: Improvements in water applications by giving only the optimum depth of irrigation water. Sprinkler and drip irrigation methods be introduced to reduce percolation losses.
- Punching underground barriers: Geological formation such as buried ridges as may be interfering with the subsoil flow leading to waterlogging conditions may be punctured to lower the water table.
- Efficient drainage system: An efficient drainage system is essential for the quick disposal of the storm water and excess irrigation water.
- Removing obstruction in natural drainage: Improved by removing debris from the water ways and providing adequate waterway under the road bridges and drainage crossings.

BIODRAINAGE

Biodrainage is defined as the process of removing the excess soil water through transpiration using bio-energy of the plant and radiation energy of the sun. It is an option to prevent the development of water logging in canal command areas. It has specific role in checking advancement of seepage from water courses and a viable option for safe disposal of large volume of saline water available in saline areas underlain with surface and sub surface drainage systems. In the absence of provision of adequate drainage, waterlogging and associated soil salinity is the major impediment to sustainability of irrigated agriculture. Though, conventional engineering drainage technologies such as subsurface or vertical drainage are able to combat the problem but are costly and generate huge quantities of drainage effluent which is difficult to dispose. Further, these options are quite successful where large tracts affected by waterlogging and salinity exist in continuity. Biodrainage, which removes the excess soil water by deep rooted fast-growing trees through evapotranspiration using bio-energy is proposed as an alternate to above engineering approaches. In rainfed areas, it can be used for interception of ground water flow and recharge control whereas in irrigated areas it is useful for controlling rise in water table, intercepting seepage water from canals and channels. Irrigation of high transpiring forest species has also been put forward for reuse of wastewater and conservation of nutrient energy into biomass and thereby bringing multiple benefits such as fuel wood production, carbon sequestration, environmental sanitation and eco-restoration. For finding an appropriate biodrainage solution, water balance between precipitation, irrigation input of water vis-à-vis water requirement of the agroecosystem and groundwater hydrology need understanding. Biodrainage potential of tree flora is a cardinal component for designing and implementing biodrainage projects. Consumptive water use of plants varies with the age, geometry, soil properties, water table, salinity and climatic conditions.

This varies between 6500 to 28000 m^3 ha^{-1} $year^{-1}$ and under ideal conditions, a tree canopy may lower water tables by 1–2 m over a time period of 3–5 years. Trees of the genes such as Eucalyptus, Populus, Casuarina, Dalbergia, Syzigium, Acacia, Prosopis, Leucaena etc. are reported effective to lowering a shallow water table and reverse salinity trends. Amongst different trees studied at different places, Eucalyptus was preferred because it grows fast in wide range of conditions, goes straight thus low shading effect on associated crops, and has luxurious water consumption in excess soil moisture conditions. Trees bio-harvest meagre amount of the salts and do not remove the salts from the soil thus may not be a good alternative when irrigation water is too saline. In cases where biodrainage results in salt accumulation, it needs to be integrated with conventional drainage technologies to make the system sustainable. Small and marginal farmers may not be able to set part of their farm aside for bio drainage activities therefore biodrainage technology may be more suitable on large farms or public lands. However, integration of trees such as Eucalyptus and Populus along with crops in a uniified agroforestry system or on approach roads or field bunds or on dykes of pond in an integrated farming system will be a viable preposition.

Introduction of canal irrigation without adequate provision for drainage in arid and semiarid regions results in water table rise leading to waterlogging and secondary salinization in large areas in irrigation commands. These problems are the result of a multitude of factors, including seepage from unlined canals, inadequate provision of surface and subsurface drainage, over irrigation and use of poor quality groundwater for irrigation. Globally about 10% of the land area is affected by waterlogging and over 6% by salinity. In India, about 6.7 million ha of land is affected by different levels and types of salinity, out of which nearly half are under irrigated agriculture. Severely waterlogged saline soils occur in about two million ha area in arid/semi- arid north western states of Punjab, Haryana, Rajasthan and Gujarat and one million ha each in the coastal and black cotton Vertisol regions of India. It is projected that about 13 million ha area in irrigation commands of India will be affected by waterlogging and soil salinity by 2025. Use of saline/alkali groundwater and climate change will further accentuate the threats. Waterlogged saline soils apart from environmental degradation result in poor crop yields by reducing crop yields by as much as 80% and finally in abandoning the land from cultivation. In India, yearly crop loss due to waterlogging has been estimated to be more than 2 m tons.

The economic loss was estimated to be about Rs. 23,900/ha with a total annual loss of Rs. 1669 million (about US$ 37 million) from the waterlogged saline area of Haryana, India. The crop yields and relative yield loss caused due to waterlogging and soil salinity of major crops are presented in table. Thus, twin menaces of waterlogging and salinization represent serious threats to the sustainability of irrigated agriculture and calls for appropriate reclamation measures.

For favourable plant growth, optimum balance of water, air and salt in the root zone is a primary requirement which can be achieved by providing adequate drainage. No doubt, the conventional technologies such as surface, subsurface or vertical drainage overcome the problem of waterlogging and salinity, but they have limitations like high cost of construction and subsequent maintenance cost and safe disposal of drainage effluent during reclamation and afterwards. Disposal of drainage effluent which contains nutrients, salts, agro-chemicals and other pollutants in water bodies had resulted in pollution of many river basins like the Indus basin in Pakistan, the Murray-Darling Basin Catchment in Australia, San Joaquin Valley in USA and various river systems in India. To combat drainage problem of such an extent, the technology needed to be less expensive, location specific, sustainable and environment friendly for its large-scale adoption by the farmers. Biodrainage could be a viable option. It is a combined drainage-cum-disposal system. Biodrainage can be defined as "pumping of excess soil water by deep-rooted vegetations through evapotranspiration using their bio-energy". Reliance on capability of vegetation to reduce water table has been reported promising both in India as well as in other countries. The main physiological feature of such vegetation is profuse transpiration whenever the root system comes in contact with groundwater. The biodrainage technique is eco-friendly as the biodrainage plantations purify the environment by absorbing greenhouse gases and releasing oxygen into the environment, environmentally safe as it does not generate any drainage effluent to dispose, economically attractive because it requires only an initial investment for planting the vegetation, and when established, the system could produce economic returns by means of fodder, wood or fiber harvested and has an additional advantage in term of carbon locked in the timber. The first documented use of the term biodrainage can be attributed to Gafni. Prior to that date Heuperman used the term bio pumping to describe the use of trees for water table control.

Table: Losses due to water logging and soil salinity.

Crop	Normal lands	Salt affected lands	Waterlogged lands
Paddy	39.9	21.8 (45)	23.0 (42)
Wheat	26.0	15.8 (40)	18.6 (38)
Cotton	16.3	6.1 (63)	3.7 (77)
Sugarcane	636.8	330.2 (48)	247.5 (61)

Waterlogging and Soil Salinity

A soil is said to be water logged when the ground water table gets linked to soil water in the crop root zone and remains like this for the remaining period in a year. It results restriction of the normal circulation of the air, decline in the level of oxygen and increase in the level of carbon dioxide. The critical depth depends on the kind of crop, but waterlogging is commonly defined as light for a soil profile depth of 3 m for substantial parts of the year, and moderate for less than 1.5 m. The severe degree occurs with a water table at 0-30 cm depth, and also included is ponding, where it rises above the surface. As an illustration, rise in water table at different places in Haryana (north India) between 1974 and 2004 is shown in figure.

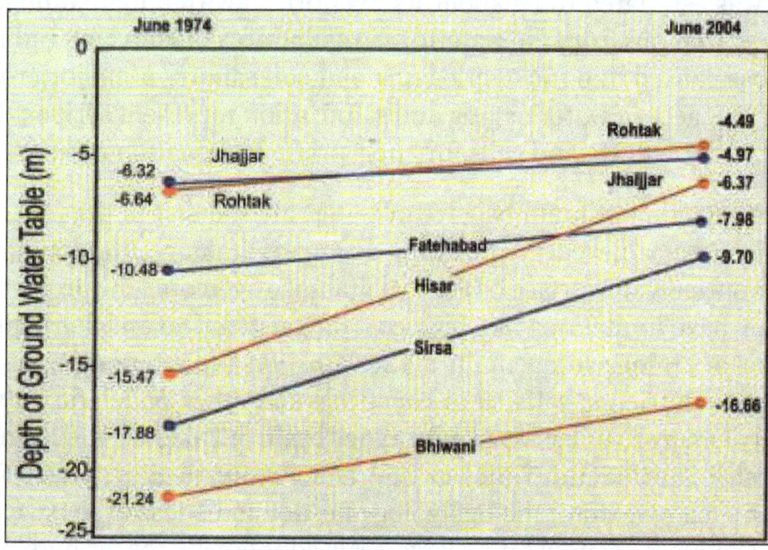

Trend in rise in water table in Haryana Soil.

Soil salinity is another problem in arid and semi arid regions. It refers to presence of excess amount of soluble salts in soil water creating its high osmotic pressure which restricts crop water uptake and ultimately resulting in poor plant growth. Salinity occurs when salts are transported to surface by rising water and got concentrated with evaporation. Several species of trees and crops can be grown in alkali, saline and saline water-logged soils. Relative tolerance of crops and trees to salt stress is reported in tables.

Table: Relative Tolerance of Crops and Grasses to Soil ESP.

Tolerant ESP, 35-50	**Moderately tolerant ESP, 15-35**	**Sensitive ESP <15**
Karnal grass (*Leptochloa fusca*)	Wheat (*Triticum aestivum*)	Gram (*Cicer arietinum*)
Rhodes grass (*Chloris gayana*)	Barley (*Hordeum vulgare*)	Mash (*Phaseolus mungo*)
Para grass (*Brachiaria mutica*)	Oat (*Avena sativa*)	Chickpea (*Cicer arietinum*)
Bermuda grass (*Cynodon dactylon*)	Shaftal (*Trifolium resupinatum*)	Lentil (*Lens esculenta*)
Rice (*Oryza sativa*)	Lucerne (*Medicago sativa*)	Soybean (*Glycine max*)
Dhaincha (*Seshania aculeate*)	Turnip (*Brassica rapa*)	Groundnut (*Arachis hypogea*)
Sugarbeet (*Beta vulgaris*)	Sunflower (*Helianthus annus*)	Sesamum (*Sesamum oriental*)
Teosinte (*Euchlaena maxicana*)	Safflower (*Carthamus tinctorius*)	Mung (*Phaseolus aureus*)
	Berseem (*Trifolium alexandrinum*)	Pea (*Pisum sativum*)
	Linseed (*Linum usitatissimum*)	Cowpea (*Vigna unguiculata*)
	Onion (*Allium cepa*)	Maize (*Zea mays*)
	Garlic (*Allium sativum*)	Cotton (*Gossypium hirsutum*)
	Pearl millet (*Pennisetum typhoides*)	

Table: Relative Tolerance of Tree Species to Soil Alkalinity.

Average pH_2 (0-120 cm)	Fuel-wood/Timber species	Fruit trees
More than 10.0	*Prosopis juliflora* (Pahari kiker) *Acacia nilotica* (Kikar) *Casuarina equisetifolia* (Australian pine)	*Achras Japota* (Chikoo)
9.0 to 10.0	*Tamarix articulata* (Frans) *Terminalia arjuna* (Arjun) *Eucalyptus tereticornis* (Safeda) *Albizzia lebbek* (Papri) *Pongamia pinnata* (Siris Karanj) *Sesbania sesban* (Dhaincha) *Emblica officinalis* (Amla)	*Zizyphus maurtiana* (Ber) *Carissa carandus* (Karaunda) *Psidium guajava* (Amrood) *Syzygium cumini* (Jamun) *Phoenix dectylifera* (Khajoor) *Aegle marmelos* (Bael)
8.2 to 9.0	*Dalbergia sisoo* (Shisham) *Morus alba* (Sehtoot) *Grevilla robusta* (Silver Oak) *Azadirachta indica* (Neem) *Tectona grandis* (Teak) *Populus deltoids* (Poplar)	*Punica granatum* (Anar) *Prunus persica* (Aru) *Pyrus communis* (Nashpati) *Vitis vinifera* (Angoor) *Mangifera Indica* (Aam)

Mechanism of Biodrainage

The root systems of trees intercept saturated zone or unsaturated capillary fringe above water table and control shallow water table. The primary objective of a bio-drainage system is to lower a shallow groundwater table to below the "critical depth" (2 m below ground surface is generally accepted as a safe depth) of the capillarity-induced evaporative processes that cause salinization. For efficient biodrainage system, trees should be fast growing having high rate transpiration system so that they absorb sufficient quantity of water from the capillary fringe located above the ground water table. The roots of herbaceous annuals have little or no contact with water table. The absorbed water is translocated to different parts of plants and finally more than 98% of the absorbed water is transpired into the atmosphere mainly through the stomata. Under ideal conditions, a tree canopy may lower water tables by 1–2 m over a time period of 3–5 years. This combined process of absorption, translocation and transpiration of excess ground water into the atmosphere by the deep-rooted vegetation conceptualizes biodrainage.

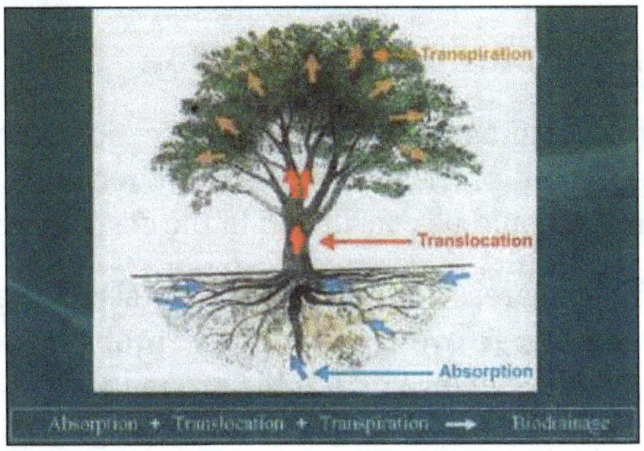

Situations Suitable for Biodrainage Application

Depending upon land use, biodrainage mechanism can be applied both in dry land/rainfed and irrigated agriculture. In rainfed systems, bio-draining ability of vegetation can be utilized mainly for recharge control, interception of ground water flow and discharge enhancement. Extensive clearing of area for agricultural development and cultivation of annual crops with shallow rooting crops may result in higher recharge and ultimately rise in water table and causing waterlogging and salinity as was observed in Kyabram, Australia. Re-vegetation of recharge areas by deep rooted tree plantation minimizes deep seepage losses in the higher parts of the landscape to minimize discharge problems referring recharge control and reduction of localized salinity and discharge problems in lower part of the landscape. Excessive evaporative demand of newly planted vegetation may lead to dry landscape causing reduced river flow, drying of well and increase in groundwater salinity. The interception of groundwater flowing through permeable layers overlying low permeability strata reduce discharge problems further down the slope. In dry land areas, extensive clearing upsets the hydrological balance, resulting in saline discharge in low lying land areas, which affects agricultural productivity and water quality in stream systems. Reclamation techniques for dry land salinity in waterlogged discharge areas focus on the restoration of the hydrological balance by planting biodrainage vegetation using the concept enhanced evapotranspiration.

In irrigated areas biodrainage can be useful for control of rise in water table, intercepting seepage water from channels and management of salinity as in convention drainage system:

- Water table control: Shallow water table causes root zone soil salinization which adversely affects crop growth. Biodrainage in irrigation areas lowers water tables below the critical depth, which is defined as the depth at which capillary salinization is negligible.
- Channel seepage interception: Channel seepage can be a major contributor to water table rise and consequently can cause water logging and salinity problems in the adjoining land. Water quality of seepage water is normally good which if intercepted can be productively used by crops. In cases if seepage water is not intercepted and left to evaporate it will increase salinity.
- Biodrainage cum conventional drainage systems: For the optimum growth of biodrainage crops also, salt accumulation in the root zone should not exceed the threshold level. Where biodrainage results in salt accumulation, engineering assistance is needed to make the system sustainable.

Planning and Design for Biodrainage Plantation

The primary requirement for planning and deciding the location for putting drainage plantings is the catchment water balance and precise identification of recharge and discharge areas in the landscape. The following issues should be considered in the development of biodrainage systems:

- Water balance: In irrigated areas due to net incremental recharge ground water table starts rising causing water logging. To overcome the above problem, the objective of any drainage scheme, is to achieve water balance before the ground water table rises up to the critical depth, which in general may be taken as 2.0 m below ground level. Plantations of deep rooted trees with high rates of transpiration should extract ground water equal or more than the net recharge so that water table is kept below the critical depth root zone. Tree plantations often use water at higher rates because of the high aerodynamic roughness which may be even more pronounced because of the socalled clothesline effect prevailing in rows of trees, substituting for a conventional drain pipe.
- Area under plantation for water balance: For sustainable water balance, the area to be covered under biodrainage vegetation should be the minimum but large enough so that amount water removed through evapotranspiration should equal the total annual recharge.

- Salt tolerance: Salt tolerance will be an important criterion for (potentially) saline discharge environments, water use considerations will prevail in recharge control situations where salinity is of no concern and in channel seepage scenarios with lowsalinity water supply. Water use capacity of trees decreases with increase in water salinity. Therefore, biodrainage crops need to be salt tolerant.

- Drawdown of water table: Crops, including trees, act as bio pumps; they depress the water table directly underneath plantation areas and consequently lower the water table in the surrounding area. Draw down effect depends on water use capacity of trees, rate of recharge in surrounding area, soil hydraulic conductivity, depth of deeper barrier layers, root system of trees and salt-tolerance of tree species.

- Salt balance: Salt balance determines the sustainability of plant water use and lowering of water table. If the salts moving into the root zone are not either (i) taken up by the vegetation and harvested or (ii) removed from the root zone by leaching, the vegetation will succumb to salinity. Large volumes of irrigation water even of low salinity significantly increase salt imports. Salinity buildup beyond a threshold level will certainly hamper plant growth. To achieve the salt level below the critical level, drainage of these salts below the crop root zone is considered a necessity for optimum plant growth. The Israeli experience has shown that the bio-drainage technique can effectively lower a shallow water table and reverse salinity trends, provided that the overall water balance is negative, i.e. that the water inputs match the water use by the tree plantation and local drainage characteristics. In absence of conventional drainage, the irrigated crops and biodrainage vegetation should be salt accumulating so that salts introduced by irrigation can be removed through crop harvest. But the ability of the biodrainage system to maintain a salt-balance is not clear. The salt uptake by plants in general is negligible compared to the total salt applied in irrigation supplies as trees do not bio-harvest as the roots exclude salts during water uptake making a saltwater lens below the root zone.

- Economic aspects: Biodrainage plantations should be of high economic value so that costs associated with planting and maintenance can be covered from the sale of the tree produce. In direct benefits of planting trees on farm lands in terms of organic carbon build-up, carbon sequestration in biomass and other eco-system services should also be considered.

Advantages

The merits of biodrainage technique over the conventional engineering-based drainage systems are as given below:

- Farmers although realize benefits of drainage but are too poor to pay cost of drainage, whereas raising biodrainage plantations is relatively less costly and affordable.

- Biodrainage requires no maintenance after initial establishment.

- No operational cost, as the plants use their bio-energy in draining out the excess ground water into atmosphere.

- Ecologically safe as drainage effluent is not produced.

- Increase in worth with age instead of depreciation.
- Preventive as well as curative system for waterlogging and salinity.
- Provides recreational areas and green open spaces, supporting beekeeping.
- Sequesters carbon and earn carbon credits.
- Moderates the temperature of the surrounding by transpiration thereby proofing for heat and cold waves.
- Mitigates green house gases by absorbing CO_2 and releasing O_2.
- Acts as wind break and protects crops in agroforestry system.
- Provides higher income to the farmers due to the production of food, fodder, timber, fuel wood and other valuable products. Thinning, based on harvesting about 50% of the slower growing trees, could provide returns even in about five years. The felled trees might be used as a source of biomass for firewood, small poles for agriculture, and/or pulp production, if markets are available. Better performing trees, could be used to produce wood for household and garden furniture.
- Biodrainage stabilizes soil on raised bed as highway avenue plantation.
- Subsurface drainage in irrigated areas is a collective activity, thus needs appropriate institutional arrangements for farmers' participation. But in case of biodrainage, there is assured people's participation as the biodrainage plantations on farmer's field belong to the individual farmers.
- The improvement in soil salinity and waterlogging provides additional land for cultivation.
- Increase in cropping intensity and soil organic carbon build-up.
- More choice among arable crops including pulses and oilseed which otherwise are sensitive to waterlogging and salinity.
- Timely sowing of crops thus facilitating better yield and profits.
- Higher crop yields and nutrient use efficiency.
- Increased employment generation and poverty reduction.

Constraints

Apart from many advantages, the following limitations of biodrainage may be kept in mind:

- Requires land, may be 10 to 15 % of the total holding of the farmer.
- Requires irrigation for the survival of the trees.
- There is a danger of damage in early stages of tree growth.

- Tree plantations may not be effective in lowering down the water table in the early growth stages.
- Competition of foliage and roots between trees and crops for light, moisture, nutrient, etc and its effects on co-existing vegetation.
- There is increased activity of the wild animals like blue bull affecting general cultivation.
- With age, there would be gradual decrease in capacities of trees for consuming and transpiring water thereby reducing extent of bio-drainage.
- In discharge sites, with evapotranspiration there can be salt accumulation in soil profile which will affect tree growth.
- Where farm holdings are small, obviously landholders are unable to set part of their farm aside for bio drainage activities. Therefore, any application of this technique will have to focus on public land.

Permissions

All chapters in this book are published with permission under the Creative Commons Attribution Share Alike License or equivalent. Every chapter published in this book has been scrutinized by our experts. Their significance has been extensively debated. The topics covered herein carry significant information for a comprehensive understanding. They may even be implemented as practical applications or may be referred to as a beginning point for further studies.

We would like to thank the editorial team for lending their expertise to make the book truly unique. They have played a crucial role in the development of this book. Without their invaluable contributions this book wouldn't have been possible. They have made vital efforts to compile up to date information on the varied aspects of this subject to make this book a valuable addition to the collection of many professionals and students.

This book was conceptualized with the vision of imparting up-to-date and integrated information in this field. To ensure the same, a matchless editorial board was set up. Every individual on the board went through rigorous rounds of assessment to prove their worth. After which they invested a large part of their time researching and compiling the most relevant data for our readers.

The editorial board has been involved in producing this book since its inception. They have spent rigorous hours researching and exploring the diverse topics which have resulted in the successful publishing of this book. They have passed on their knowledge of decades through this book. To expedite this challenging task, the publisher supported the team at every step. A small team of assistant editors was also appointed to further simplify the editing procedure and attain best results for the readers.

Apart from the editorial board, the designing team has also invested a significant amount of their time in understanding the subject and creating the most relevant covers. They scrutinized every image to scout for the most suitable representation of the subject and create an appropriate cover for the book.

The publishing team has been an ardent support to the editorial, designing and production team. Their endless efforts to recruit the best for this project, has resulted in the accomplishment of this book. They are a veteran in the field of academics and their pool of knowledge is as vast as their experience in printing. Their expertise and guidance has proved useful at every step. Their uncompromising quality standards have made this book an exceptional effort. Their encouragement from time to time has been an inspiration for everyone.

The publisher and the editorial board hope that this book will prove to be a valuable piece of knowledge for students, practitioners and scholars across the globe.

Index

www.ingramcontent.com/pod-product-compliance
Lightning Source LLC
LaVergne TN
LVHW080659210926
848485LV00003B/398